Advance Praise for *The Magic of Math*

"They say magicians should never reveal their secrets. Happily, Arthur Benjamin has ignored this silly adage—for in this small volume, Benjamin reveals to his audience the secrets of numbers and other mathematical illusions that have intrigued mathematicians for millennia."

—Edward B. Burger, president, Southwestern University, and author of *The 5 Elements of Effective Thinking*

"This book will be magical for my students, as it would have been for me throughout my school days. They'll be able to revisit the book frequently as they learn more math, finding deeper appreciation and discovering new areas to explore with each visit."

—Richard Rusczyk, founder, Art of Problem Solving, and director, USA Mathematical Talent Search

"In *The Magic of Math*, Arthur Benjamin has pulled off a seemingly impossible trick. He has made higher mathematics appear so natural and engaging that you will wonder why you were ever bored and confused in math class. There are many books that attempt to popularize mathematics. This is one of the best. On virtually every page I found myself learning new things, or looking at familiar topics in novel ways."

—Jason Rosenhouse, professor of Mathematics, James Madison University, and author of *The Monty Hall Problem*

"In *The Magic of Math*, mathemagician Arthur Benjamin gives us an entertaining and enlightening tour of a wide swath of fundamental mathematical ideas, presented in a way that is accessible to a broad audience. A particularly appealing feature of the book is the frequent use of friendly, down-to-earth explanations of the concepts and connections between them."

—Ronald Graham, president emeritus, American Mathematical Society, and coauthor of *Magical Mathematics*

"This book is a whirlwind tour of mathematics from arithmetic and algebra all the way to calculus and infinity, and especially the number 9. Arthur Benjamin's enthusiastic and engaging writing style makes *The Magic of Math* a great addition to any math enthusiast's bag of tricks."

—Laura Taalman, professor of Mathematics and Statistics, James Madison University

"Mathematics is full of surprisingly beautiful patterns, which Arthur Benjamin's witty personality brings to life in *The Magic of Math*. You will not only discover many wonderful ideas, but you will also find some fun mathematical magic tricks that you will want to try out on your friends and family. Be prepared to learn that math is more entertaining than you may have thought."

—George W. Hart, mathematical sculptor, research professor, Stony Brook University, and cofounder, The Museum of Mathematics

"*The Magic of Math* is a delightful stroll through a garden filled with fascinating examples. Anyone with any interest in magic, puzzles, or math will have many hours of enjoyment in reading this book."

—Maria M. Klawe, president, Harvey Mudd College

"Arthur Benjamin has created an instant mathematical classic, by combining Isaac Asimov's clarity with Martin Gardner's taste and adding his own sense of fun and adventure. I wish he wrote this book when I was a kid."

—Paul A. Zeitz, professor and chair of Mathematics, University of San Francisco, and author of *The Art and Craft of Problem Solving*

"There's a playful joy to be found in this book, for readers at any level. Most magicians don't reveal their secrets, but in *The Magic of Math*, Arthur Benjamin shows how uncovering the mystery behind beautiful mathematical truths makes math even more marvelous to behold."

—Francis Su, president, Mathematical Association of America

"*The Magic of Math* offers an expansive, unforgettable journey through mathematics where numbers dance and mathematical secrets are revealed. Just open the book and start reading; you'll be swept over by the magic of Benjamin's writing. Luckily, there is no magician's code to these secrets as you'll undoubtedly want to share and perform them with family and friends."

—Tim Chartier, professor of Mathematics, Davidson College, and author of *Math Bytes*

THE
MAGIC
OF MATH

THE
MAGIC
OF MATH

Solving for x and Figuring Out Why

ARTHUR BENJAMIN

BASIC BOOKS
A Member of the Perseus Books Group
New York

Illustrations by Natalya St. Clair

A catalog record for this book is available from the Library of Congress.

Library of Congress Control Number: 2015936185
ISBN: 978-0-465-05472-5 (hardcover)
ISBN: 978-0-465-06162-4 (e-book)

10 9 8 7 6 5 4 3

I dedicate this book to my wife, Deena,
and daughters, Laurel and Ariel.

Contents

Intr0ducti0n

Throughout my life, I have always had a passion for magic. Whether I was watching other magicians or performing magic myself, I was fascinated with the methods used to accomplish amazing and impressive feats, and I loved learning its secrets. With just a handful of simple principles, I could even invent tricks of my own.

I had the same experience with mathematics. From a very early age, I saw that numbers had a magic all their own. Here's a trick you might enjoy. Think of a number between 20 and 100. Got it? Now add your digits together. Now subtract the total from your original number. Finally, add the digits of the new number together. Are you thinking of the number 9? (If not, you might want to check your previous calculation.) Pretty cool, huh? Mathematics is filled with magic like this, but most of us are never exposed to it in school. In this book, you will see how numbers, shapes, and pure logic can yield delightful surprises. And with just a little bit of algebra or geometry, you can often discover the secrets behind the magic, and perhaps even discover some beautiful mathematics of your own.

This book covers the essential mathematical subjects like numbers, algebra, geometry, trigonometry, and calculus, but it also covers topics that are not so well represented, like Pascal's triangle, infinity, and magical properties of numbers like 9, π, e, i, Fibonacci numbers, and the golden ratio. And although none of the big mathematical subjects can be completely covered in just a few dozen pages, I hope you come away with an understanding of the major concepts, a better idea of why they work, and an appreciation of the elegance and relevance of each subject. Even if you have seen some of these topics before, I hope you will see them and enjoy them with new perspectives. And as we learn more mathematics, the magic becomes more sophisticated and fascinating. For example, here is one of my favorite equations:

$$e^{i\pi} + 1 = 0$$

Some refer to this as "God's equation," because it uses the most important numbers in mathematics in one magical equation. Specifically, it uses 0 and 1, which are the foundations of arithmetic; $\pi = 3.14159\ldots$, which is the most important number in geometry; $e = 2.71828\ldots$, which is the most important number in calculus; and the *imaginary* number i, with a square of -1. We'll say more about π in Chapter 8, and the numbers i and e are described in greater detail in Chapter 10. In Chapter 11, we'll see the mathematics that help us understand this magical equation.

My target audience for this book is anyone who will someday need to take a math course, is currently taking a math course, or is finished taking math courses. In other words, I want this book to be enjoyed by everyone, from math-phobics to math-lovers. In order to do this, I need to establish some rules.

> ✗ **Rule 1: You can skip the gray boxes (except this one)!**
> Each chapter is filled with "Asides," where I like to go off on a tangent to talk about something interesting. It might be an extra example or a proof, or something that will appeal to the more advanced readers. You might want to skip these the first time you read this book (and maybe the second and third times too). And I do hope that you reread this book. Mathematics is a subject that is worth revisiting.

Rule 2: Don't be afraid to skip paragraphs, sections, or even chapters. In addition to skipping the gray boxes, feel free to go forward anytime you get stuck. Sometimes you need perspective on a topic before it fully sinks in. You will be surprised how much easier a topic can be when you come back to it later. It would be a shame to stop partway through the book and miss all the fun stuff that comes later.

Rule 3: Don't skip the last chapter. The last chapter, on the mathematics of infinity, has lots of mind-blowing ideas that they probably won't teach you in school, and many of these results do not rely on the earlier chapters. On the other hand, the last chapter does refer to ideas that appear in all of the previous chapters, so that might give you the extra incentive to go back and reread previous parts of the book.

Rule π: Expect the unexpected. While mathematics is a seriously important subject, it doesn't have to be taught in a serious and dry fashion. As a professor of mathematics at Harvey Mudd College, I can't resist the occasional pun, joke, poem, song, or magic trick to make a class more enjoyable, and they appear throughout these pages. And since this is a book, you don't have to hear me sing — lucky for you!

Follow these rules, and discover the magic of mathematics!

CHAPTER ONE

$1 + 2 + 3 + 4 + \cdots + 100 = 5050$

The Magic of Numbers

Number Patterns

The study of mathematics begins with numbers. In school, after we learn how to count and represent numbers using words or digits or physical objects, we spend many years manipulating numbers through addition, subtraction, multiplication, division, and other arithmetical procedures. And yet, we often don't get to see that numbers possess a magic of their own, capable of entertaining us, if we just look below the surface.

Let's start with a problem given to a mathematician named Karl Friedrich Gauss when he was just a boy. Gauss's teacher asked him and his classmates to add up all the numbers from 1 to 100, a tedious task designed to keep the students busy while the teacher did other work. Gauss astonished his teacher and classmates by immediately writing down the answer: 5050. How did he do it? Gauss imagined the numbers 1 through 100 split into two rows, with the numbers 1 through 50 on the top and the numbers 51 through 100 *written backward* on the bottom, as shown below. Gauss observed that each of the 50 columns would add up to the same sum, 101, and so their total would just be 50×101, which is 5050.

1	2	3	4	...	47	48	49	50
+ 100	+ 99	+ 98	+ 97	...	+ 54	+ 53	+ 52	+ 51
101	101	101	101	...	101	101	101	101

Splitting the numbers from 1 to 100 into two rows; each pair of numbers adds to 101

Gauss went on to become the greatest mathematician of the nineteenth century, not because he was quick at doing mental calculations, but because of his ability to make numbers dance. In this chapter, we will explore many interesting number patterns and start to see how numbers dance. Some of these patterns can be applied to do mental calculations more quickly, and some are just beautiful for their own sake.

We've used Gauss's logic to sum the first 100 numbers, but what if we wanted to sum 17 or 1000 or 1 million? We will, in fact, use his logic to sum the first n numbers, where n can be any number you want! Some people find numbers to be less abstract when they can visualize them. We call the numbers 1, 3, 6, 10, and 15 *triangular numbers*, since we can create triangles like the ones below using those quantities of dots. (You might dispute that 1 dot forms a triangle, but nevertheless 1 is considered triangular.) The official definition is that the nth triangular number is $1 + 2 + 3 + \cdots + n$.

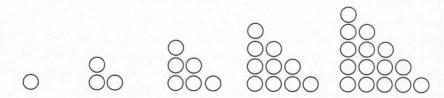

The first 5 triangular numbers are 1, 3, 6, 10, and 15

Notice what happens when we put two triangles side by side, as depicted on the opposite page:

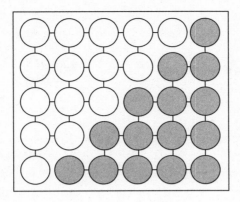

How many dots are in the rectangle?

Since the two triangles form a rectangle with 5 rows and 6 columns, there are 30 dots altogether. Hence, each original triangle must have half as many dots, namely 15. Of course, we knew that already, but the same argument shows that if you take two triangles with n rows and put them together as we did, then you form a rectangle with n rows and $n + 1$ columns, which has $n \times (n + 1)$ dots (often written more succinctly as $n(n + 1)$ dots). As a result, we have derived the promised formula for **the sum of the first n numbers:**

$$1 + 2 + 3 + \cdots + n = \frac{n(n + 1)}{2}$$

Notice what we just did: we saw a pattern to sum the first 100 numbers and were able to extend it to handle any problem of the same form. If we needed to add the numbers 1 through 1 million, we could do it in just two steps: multiply 1,000,000 by 1,000,001, then divide by 2!

Once you figure out one mathematical formula, other formulas often present themselves. For example, if we double both sides of the last equation, we get a formula for **the sum of the first n even numbers:**

$$2 + 4 + 6 + \cdots + 2n = n(n + 1)$$

What about **the sum of the first n odd numbers?** Let's look at what the numbers seem to be telling us.

$$
\begin{aligned}
1 &= 1 \\
1 + 3 &= 4 \\
1 + 3 + 5 &= 9 \\
1 + 3 + 5 + 7 &= 16 \\
1 + 3 + 5 + 7 + 9 &= 25 \\
&\vdots
\end{aligned}
$$

What is the sum of the first n odd numbers?

The numbers on the right are *perfect squares*: $1 \times 1, 2 \times 2, 3 \times 3$, and so on. It's hard to resist noticing the pattern that the sum of the first n odd numbers seems to be $n \times n$, often written as n^2. But how can we be sure that this is not just some temporary coincidence? We'll see a few ways to derive this formula in Chapter 6, but such a simple pattern should have a simple explanation. My favorite justification uses a count-the-dots strategy again, and reminds us of why we call numbers like 25 perfect *squares*. Why should the first 5 odd numbers add to 5^2? Just look at the picture of the 5-by-5 square below.

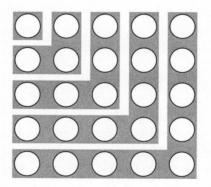

How many dots are in the square?

This square has $5 \times 5 = 25$ dots, but let's count the dots another way. Start with the 1 dot in the upper left corner. It is surrounded by 3 dots, then 5 dots, then 7 dots, then 9 dots. Consequently,

$$1 + 3 + 5 + 7 + 9 = 5^2$$

If we started with an n-by-n square, then we can break it into n (back-ward) L-shaped regions of sizes $1, 3, 5, \ldots, (2n - 1)$. When viewed this way, we have a formula for **the sum of the first n odd numbers:**

$$1 + 3 + 5 + \cdots + (2n - 1) = n^2$$

> ✂ **Aside**
>
> Later in this book, we'll see how the approach of counting dots (and the general approach of answering a question in two different ways) leads to some interesting results in advanced mathematics. But it can also be useful for understanding elementary mathematics as well. For example, *why does* $3 \times 5 = 5 \times 3$? I'm sure you haven't even questioned that statement since you were told, as a child, that the order of multiplication doesn't matter. (Mathematicians say that multiplication of numbers is *commutative*.) But why should 3 bags of 5 marbles contain the same amount as 5 bags of 3 marbles? The explanation is simple if you just count the dots in a 3-by-5 rectangle. Counting row by row, we see 3 rows of 5 dots apiece, giving us 3×5 dots. On the other hand, we also have 5 columns with 3 dots apiece, so there are also 5×3 dots.
>
>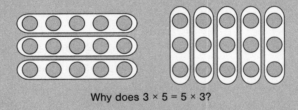
>
> Why does $3 \times 5 = 5 \times 3$?

Let's apply the pattern from the sum of odd numbers to find an even more beautiful pattern. If our goal is to make the numbers dance, then you might say we are about to do some *square dancing*.

Consider this interesting pyramid of equations:

$$1 + 2 = 3$$
$$4 + 5 + 6 = 7 + 8$$
$$9 + 10 + 11 + 12 = 13 + 14 + 15$$
$$16 + 17 + 18 + 19 + 20 = 21 + 22 + 23 + 24$$
$$25 + 26 + 27 + 28 + 29 + 30 = 31 + 32 + 33 + 34 + 35$$
$$\vdots$$

What patterns do you see? It's easy to count the numbers in each row: $3, 5, 7, 9, 11$, and so on. Next comes an unexpected pattern. What is the first number of each row? Judging from the first 5 rows, 1, 4, 9, 16, 25, ... , they appear to be the perfect squares. Why is that? Let's look at the fifth row. How many numbers appear before row 5? If we count the numbers in the preceding four rows, we have $3 + 5 + 7 + 9$. To get the leading number of row 5, we just add 1 to this sum, so we really have the sum of the first 5 odd numbers, which we now know to be 5^2.

Now let's verify the fifth equation without actually adding any numbers. What would Gauss do? If we temporarily ignore the 25 at the beginning of the row, then there are 5 remaining numbers on the left, which are each 5 less than their corresponding numbers on the right.

25	26	27	28	29	30
	$-\ 31$	$-\ 32$	$-\ 33$	$-\ 34$	$-\ 35$
	-5	-5	-5	-5	-5

Comparing the left side of row 5 with the right side of row 5

Hence the five numbers on the right have a total that is 25 greater than their corresponding numbers on the left. But this is compensated for by the number 25 on the left. Hence the sums balance as promised. By the same logic, and a little bit of algebra, it can be shown that this pattern will continue indefinitely.

> ## ✂ Aside
>
> For those who wish to see the little bit of algebra now, here it is. Row n is preceded by $3 + 5 + 7 + \cdots + (2n - 1) = n^2 - 1$ numbers, so the left side of the equation must start with the number n^2, followed by the next n consecutive numbers, $n^2 + 1$ through $n^2 + n$. The right side has n consecutive numbers starting with $n^2 + n + 1$ through $n^2 + 2n$. If we temporarily ignore the n^2 number on the left, we see that the n numbers on the right are each n larger than their corresponding numbers on the left, so their difference is $n \times n$, which is n^2. But this is compensated for on the left by the initial n^2 term, so the equations balance.

Time for a new pattern. We saw that odd numbers could be used to make squares. Now let's see what happens when we put all the odd numbers in one big triangle, as shown on the next page.

We see that $3 + 5 = 8, 7 + 9 + 11 = 27, 13 + 15 + 17 + 19 = 64$. What do the numbers 1, 8, 27, and 64 have in common? They are perfect cubes! For example, summing the five numbers in the fifth row, we get

$$
\begin{array}{ccccccccc}
& & & & 1 & & & = & 1 & = & 1^3 \\
& & & 3 & + & 5 & & = & 8 & = & 2^3 \\
& & 7 & + & 9 & + & 11 & = & 27 & = & 3^3 \\
& 13 & + & 15 & + & 17 & + & 19 & = 64 & = & 4^3 \\
21 & + & 23 & + & 25 & + & 27 & + & 29 & = 125 & = & 5^3
\end{array}
$$

$$\vdots \qquad\qquad \vdots \qquad \vdots$$

An odd triangle

$$21 + 23 + 25 + 27 + 29 = 125 = 5 \times 5 \times 5 = 5^3$$

The pattern seems to suggest that the sum of the numbers in the nth row is n^3. Will this always be the case, or is it just some *odd* coincidence? To help us understand this pattern, check out the middle numbers in rows 1, 3, and 5. What do you see? The perfect squares 1, 9, and 25. Rows 2 and 4 don't have middle numbers, but surrounding the middle are the numbers 3 and 5 with an average of 4, and the numbers 15 and 17 with an average of 16. Let's see how we can exploit this pattern.

Look again at row 5. Notice that we can *see* that the sum is 5^3 without actually adding the numbers by noticing that these five numbers are symmetrically centered around the number 25. Since the average of these five numbers is 5^2, then their total must be $5^2 + 5^2 + 5^2 + 5^2 + 5^2 = 5 \times 5^2$, which is 5^3. Similarly, the average of the four numbers of row 4 is 4^2, so their total must be 4^3. With a little bit of algebra (which we won't do here), you can show that the average of the n numbers in row n is n^2, so their total must be n^3, as desired.

Since we're talking about cubes and squares, I can't resist showing you one more pattern. What totals do you get as you add the cubes of numbers starting with 1^3?

$$
\begin{aligned}
1^3 &= 1 = \mathbf{1}^2 \\
1^3 + 2^3 &= 9 = \mathbf{3}^2 \\
1^3 + 2^3 + 3^3 &= 36 = \mathbf{6}^2 \\
1^3 + 2^3 + 3^3 + 4^3 &= 100 = \mathbf{10}^2 \\
1^3 + 2^3 + 3^3 + 4^3 + 5^3 &= 225 = \mathbf{15}^2
\end{aligned}
$$

$$\vdots$$

The sum of the cubes is always a perfect square

When we start summing cubes, we get the totals $1, 9, 36, 100, 225$, and so on, which are all perfect squares. But they're not just *any* perfect squares; they are the squares of $1, 3, 6, 10, 15$, and so on, which are all triangular numbers! Earlier we saw that these were the sums of integers and so, for example,

$$1^3 + 2^3 + 3^3 + 4^3 + 5^3 = 225 = 15^2 = (1 + 2 + 3 + 4 + 5)^2$$

To put it another way, the sum of the cubes of the first n numbers is the square of the sum of the first n numbers. We're not quite ready to prove that result now, but we will see two proofs of this in Chapter 6.

Fast Mental Calculations

Some people look at these number patterns and say, "Okay, that's nice. But what good are they?" Most mathematicians would probably respond like any artist would—by saying that a beautiful pattern needs no justification other than its beauty. And the patterns become even more beautiful the more deeply we understand them. But sometimes the patterns can lead to real applications.

Here's a simple pattern that I had the pleasure of discovering (even if I wasn't the first person to do so) when I was young. I was looking at pairs of numbers that added up to 20 (such as 10 and 10, or 9 and 11), and I wondered how large the product could get. It seemed that the largest product would occur when both numbers were equal to 10, and the pattern confirmed that.

			Distance Below 100
10×10	$=$	100	
9×11	$=$	99	1
8×12	$=$	96	4
7×13	$=$	91	9
6×14	$=$	84	16
5×15	$=$	75	25
	$\vdots$		$\vdots$

The product of numbers that add to 20

The pattern was unmistakable. As the numbers were pulled farther apart, the product became smaller. And how far below 100 were they?

$1, 4, 9, 16, 25, \ldots$, which were $1^2, 2^2, 3^2, 4^2, 5^2$, and so on. Does this pattern always work? I decided to try another example, by looking at pairs of numbers that add up to 26.

<u>Distance Below 169</u>

13 × 13 = 169	
12 × 14 = 168	1
11 × 15 = 165	4
10 × 16 = 160	9
9 × 17 = 153	16
8 × 18 = 144	25
⋮	⋮

The product of numbers that add to 26

Once again, the product was maximized when we chose the two numbers to be equal, and then the product decreased from 169 by 1, then 4, then 9, and so on. After a few more examples, I was convinced that the pattern was true. (I'll show you the algebra behind it later.) Then I saw a way that this pattern could be applied to squaring numbers faster.

Suppose we want to square the number 13. Instead of performing 13×13 directly, we will perform the easier calculation of $10 \times 16 = 160$. This is almost the answer, but since we went up and down 3, it is shy of the answer by 3^2. Thus,

$$13^2 = (10 \times 16) + 3^2 = 160 + 9 = 169$$

Let's try another example. Try doing 98×98 using this method. To do this, we go up 2 to 100, then down 2 to 96, then add 2^2. That is,

$$98^2 = (100 \times 96) + 2^2 = 9600 + 4 = 9604$$

Squaring numbers that end in 5 are especially easy, since when you go up and down 5, the numbers you are multiplying will both end in 0. For example,

$$35^2 = (30 \times 40) + 5^2 = 1200 + 25 = 1225$$

$$55^2 = (50 \times 60) + 5^2 = 3000 + 25 = 3025$$

$$85^2 = (80 \times 90) + 5^2 = 7200 + 25 = 7225$$

Now try 59^2. By going up and down 1, you get $59^2 = (60 \times 58) + 1^2$ But how should you mentally calculate 60×58? Three words of advice: left to right. Let's first ignore the 0 and compute 6×58 from left to right. Now $6 \times 50 = 300$ and $6 \times 8 = 48$. Add those numbers together (from left to right) to get 348. Therefore, $60 \times 58 = 3480$, and so

$$59^2 = (60 \times 58) + 1^2 = 3480 + 1 = 3481$$

✂ Aside

Here's the algebra that explains why this method works. (You may want to come back to this after reading about *the difference of squares* in Chapter 2.)

$$A^2 = (A + d)(A - d) + d^2$$

where A is the number being squared, and d is the distance to the nearest easy number (although the formula works for any number d). For example, when squaring 59, $A = 59$ and $d = 1$, so the formula tells you to do $(59 + 1) \times (59 - 1) + 1^2$, as in the previous calculation.

Once you get good at squaring two-digit numbers, you can square three-digit numbers by the same method. For example, if you know that $12^2 = 144$, then

$$112^2 = (100 \times 124) + 12^2 = 12{,}400 + 144 = 12{,}544$$

A similar method can be used for multiplying any two numbers that are close to 100. When you first see the method, it looks like pure magic. Look at the problem 104×109. Next to each number we write down its distance from 100, as in the figure below. Now add the first number to the second distance number. Here, that would be $104 + 9 = 113$. Then multiply the distance numbers together. In this case, $4 \times 9 = 36$. Push those numbers together and your answer magically appears.

$$
\begin{array}{r}
104 \ (4) \\
\times \ \ 109 \ (9) \\
\hline
113 \ \ 36
\end{array}
$$

A magical way to multiply numbers close to 100—here, $104 \times 109 = 11{,}336$

I'll show you more examples of this and the algebra behind it in Chapter 2. But while we're on the subject, let me say a few more words about mental math. We spend an awful amount of time learning pencil-and-paper arithmetic, but precious little time learning how to do math in your head. And yet, in most practical situations, you are more likely to need to calculate mentally than to calculate on paper. For most large calculations, you will use a calculator to get the exact answer, but you generally don't use a calculator when reading a nutrition label or hearing a speech or listening to a sales report. For those situations, you typically just want a good mental estimate of the important quantities. The methods taught in school are fine for doing math on paper, but they are generally poor for doing math in your head.

I could write a book on fast mental math strategies, but here are some of the essential ideas. The main tip, which I cannot emphasize enough, is to do the problems from *left to right*. Mental math is a process of constant simplification. You start with a hard problem and simplify it to easier problems until you reach your answer at the end.

Mental addition.

Consider a problem like

$$314 + 159$$

(I'm writing the numbers horizontally so you're less tempted to go into pencil-and-paper mode.) Starting with 314, first add the number 100 to give us a simpler addition problem:

$$414 + 59$$

Adding 50 to 414 gives us an even simpler problem that we can solve right away:

$$464 + 9 = 473$$

That is the essence of mental addition. The only other occasionally useful strategy is that sometimes we can turn a hard addition problem into an easy subtraction problem. This often arises if we are adding the price of retail items. For example, let's do

$$\$23.58 + \$8.95$$

Since $8.95 is 5 cents below $9, we first add $9 to $23.58, then subtract 5 cents. The problem simplifies to

$$\$32.58 - \$0.05 = \$32.53$$

Mental subtraction.

The most important idea with mental subtraction problems is the strategy of *oversubtracting*. For example, when subtracting 9, it's often easier to first subtract 10, then add back 1. For example,

$$83 - 9 = 73 + 1 = 74$$

Or to subtract 39, it's probably easier to first subtract 40, then add back 1.

$$83 - 39 = 43 + 1 = 44$$

When subtracting numbers with two or more digits, the key idea is to use *complements*. (You'll compliment me later for this.) The complement of a number is its distance to the next-highest round number. With one-digit numbers, this is the distance to 10. (For example, the complement of 9 is 1.) For two-digit numbers, this is the distance to 100. Look at the following pairs of numbers that add to 100. What do you notice?

$$
\begin{array}{ccccc}
87 & 75 & 56 & 92 & 80 \\
+\ 13 & +\ 25 & +\ 44 & +\ 08 & +\ 20 \\
\hline
100 & 100 & 100 & 100 & 100 \\
\end{array}
$$

Complementary two-digit numbers sum to 100

We say that the complement of 87 is 13, the complement of 75 is 25, and so on. Conversely, the complement of 13 is 87 and the complement of 25 is 75. Reading each problem from left to right, you will notice that (except for the last problem) the leftmost digits add to 9 and the rightmost digits add to 10. The exception is when the numbers end in 0 (as in the last problem). For example, the complement of 80 is 20.

Let's apply the strategy of complements for the problem $1234 - 567$. Now, that would not be a fun problem to do on paper. But with complements, *hard subtraction problems become easy addition problems!* To subtract 567, we begin by subtracting 600. That's easy to do, especially if you think from left to right: $1234 - 600 = 634$. But you've subtracted too much. How much too much? Well, how far is 567 from 600? It's the same as the distance between 67 and 100, which is 33. Thus

$$1234 - 567 = 634 + 33 = 667$$

Notice that the addition problem is especially easy because there are no "carries" involved. This will often be the case when doing subtraction problems by complements:

Something similar happens with three-digit complements.

$$
\begin{array}{r} 789 \\ +\ \ 211 \\ \hline 1000 \end{array}
\qquad
\begin{array}{r} 555 \\ +\ \ 445 \\ \hline 1000 \end{array}
\qquad
\begin{array}{r} 870 \\ +\ \ 130 \\ \hline 1000 \end{array}
$$

Complementary three-digit numbers sum to 1000

For most problems (when the number does not end in zero), the corresponding digits sum to 9, except the last pair of digits sum to 10. For example, with 789, $7 + 2 = 9$, $8 + 1 = 9$, and $9 + 1 = 10$. This can be handy when making change. For example, my favorite sandwich from my local deli costs \$6.76. How much change would I get from \$10.00? The answer is found by taking the complement of 676, which is 324. Hence the change back is \$3.24.

> ✗ **Aside**
>
> Whenever I buy this sandwich, I can't help but notice that both the price and the change were perfect squares ($26^2 = 676$ and $18^2 = 324$). (Bonus question: There is another pair of perfect squares that add up to 1000. Can you find them?)

Mental multiplication.

After you have memorized your multiplication table through 10, you can mentally calculate, at least approximately, the answer to any multiplication problem. The next step is to master (but not memorize!) your one-digit times two-digit multiplication problems. The key idea is to work from left to right. For example, when multiplying 8×24, you should first multiply 8×20, then add this to 8×4:

$$8 \times 24 = (8 \times 20) + (8 \times 4) = 160 + 32 = 192$$

Once you've mastered those, it's time to practice one-digit times three-digit multiplication problems. These are a bit trickier, since there is more to keep in your memory. The key here is to gradually add the numbers as you go along so there is not as much to remember. For example, when multiplying 456×7, you stop to add $2800 + 350$, as below, before adding 42 to it.

$$
\begin{array}{r}
456 \\
\times \quad 7 \\
\hline
400 \times 7 = \quad 2800 \\
50 \times 7 = + \quad 350 \\
\hline
3150 \\
6 \times 7 = + \quad 42 \\
\hline
3192 \\
\end{array}
$$

Once you have the hang of doing problems of this size, then it's time to move on to two-digit times two-digit problems. For me, this is where the fun begins, because there are usually many different ways you can attack these problems. By doing the problem multiple ways, you can check your answer—and simultaneously revel in the consistency of arithmetic! I'll illustrate all of these methods with a single example, 32×38.

The most familiar method (most closely resembling what you do on paper) is the *addition method*, which can be applied to any problem. Here we break up one number (usually the one with the smaller ones digit) into two parts, then multiply each part by the other number, and add the results together. For example,

$$32 \times 38 = (30 + 2) \times 38 = (30 \times 38) + (2 \times 38) = \cdots$$

Now how do we calculate 30×38? Let's do 3×38, then attach the 0 at the end. Now $3 \times 38 = 90 + 24 = 114$, so $30 \times 38 = 1140$. Then $2 \times 38 = 60 + 16 = 76$, so

$$32 \times 38 = (30 \times 38) + (2 \times 38) = 1140 + 76 = 1216$$

Another way to do a problem like this (typically when one of the numbers ends in 7, 8, or 9) is to use the *subtraction method*. Here we exploit the fact that $38 = 40 - 2$ to get

$$38 \times 32 = (40 \times 32) - (2 \times 32) = 1280 - 64 = 1216$$

The challenge with the addition and subtraction methods is that they require you to hold on to a big number (like 1140 or 1280) while doing a separate calculation. That can be difficult. My usual preferred method for two-digit multiplication is the *factoring method*, which can be applied whenever one of the numbers can be expressed as the product of two 1-digit numbers. In our example, we see that 32 can be factored as 8×4. Consequently,

$$38 \times 32 = 38 \times 8 \times 4 = 304 \times 4 = 1216$$

If we factor 32 as 4×8, we get $38 \times 4 \times 8 = 152 \times 8 = 1216$, but I prefer to multiply the two-digit number by the larger factor first, so that the next number (usually a three-digit number) is multiplied by the smaller factor.

> ## ✂ Aside
>
> The factoring method also works well on multiples of 11, since there is an especially easy trick for multiplying by 11: *just add the digits and put the total in between.* For example, to do 53×11, we see that $5 + 3 = 8$, so the answer is 583. What's 27×11? Since $2 + 7 = 9$, the answer is 297. What if the total of the two digits is bigger than 9? In that case, we insert the last digit of the total and increase the first digit by 1. For example, to compute 48×11, since $4 + 8 = 12$, the answer is 528. Similarly, $74 \times 11 = 814$. This can be exploited when multiplying numbers by multiples of 11. For example,
>
> $$74 \times 33 = 74 \times 11 \times 3 = 814 \times 3 = 2442$$

Another fun method for multiplying two-digit numbers is the *close together method*. You can use it when *both numbers begin with the same digit*. It seems utterly magical when you first watch it in action. For example, would you believe that

$$38 \times 32 = (30 \times 40) + (8 \times 2) = 1200 + 16 = 1216$$

The calculation is especially simple (as in the example above) when the second digits sum to 10. (Here, both numbers begin with 3 and the second digits have the sum $8 + 2 = 10$.) Here's another example:

$$83 \times 87 = (80 \times 90) + (3 \times 7) = 7200 + 21 = 7221$$

Even when the second digits don't add up to 10, the calculation is almost as simple. For example, to multiply 41×44, if you decrease the smaller number by 1 (to reach the round number 40), then you must increase the larger number by 1 as well. Consequently,

$$41 \times 44 = (40 \times 45) + (1 \times 4) = 1800 + 4 = 1804$$

For 34×37, if you decrease 34 by 4 (to reach the round number 30), then it gets multiplied by $37 + 4 = 41$, and then we add 4×7 as follows:

$$34 \times 37 = (30 \times 41) + (4 \times 7) = 1230 + 28 = 1258$$

By the way, the mysterious multiplication we saw earlier of 104×109 was just an application of this same method.

$$104 \times 109 = (100 \times 113) + (04 \times 09) = 11300 + 36 = 11{,}336$$

Some schools are asking students to memorize their multiplication tables through 20. Rather than memorize these products, we can calculate them quickly enough using this method. For example,

$$17 \times 18 = (10 \times 25) + (7 \times 8) = 250 + 56 = 306$$

Why does this mysterious method work? For this, we'll need algebra, which we will discuss in Chapter 2. And once we have algebra, we can find new ways to calculate. For example, we'll see why the last problem can also be done as follows:

$$18 \times 17 = (20 \times 15) + ((-2) \times (-3)) = 300 + 6 = 306$$

Speaking of the multiplication table, check out the one-digit table on the opposite page that I promised earlier. Here's a question that would appeal to a young Gauss: *What is the sum of all the numbers in the multiplication table?* Take a minute and see if you can figure it out in an elegant way. I'll provide the answer at the end of the chapter.

Mental estimation and division.

Let's begin with a very simple question with a very simple answer that we are rarely taught in school:

(a) If you multiply two 3-digit numbers together, can you immediately tell how many digits can be in the answer?

And a follow-up question:

(b) How many digits can be in the answer when multiplying a four-digit number by a five-digit number?

We spend so much time in school learning to generate the digits of a multiplication or division problem, and very little time thinking about the important aspects of the answer. Yet it's way more important to know the approximate *size* of the answer than to know the last

×	1	2	3	4	5	6	7	8	9	10
1	1	2	3	4	5	6	7	8	9	10
2	2	4	6	8	10	12	14	16	18	20
3	3	6	9	12	15	18	21	24	27	30
4	4	8	12	16	20	24	28	32	36	40
5	5	10	15	20	25	30	35	40	45	50
6	6	12	18	24	30	36	42	48	54	60
7	7	14	21	28	35	42	49	56	63	70
8	8	16	24	32	40	48	56	64	72	80
9	9	18	27	36	45	54	63	72	81	90
10	10	20	30	40	50	60	70	80	90	100

What is the sum of all 100 numbers in the multiplication table?

digits or even the first digits. (Knowing that the answer begins with 3 is meaningless until you know whether the answer will be closer to 30,000 or 300,000 or 3,000,000.) The answer to question (a) is five or six digits. Why is that? The smallest possible answer is $100 \times 100 = 10,000$, which has five digits. The biggest possible answer is 999×999, which is strictly less than $1000 \times 1000 = 1,000,000$, which has seven digits (but just barely!). Since 999×999 is smaller, then it must have six digits. (Of course, you could easily compute the last answer in your head: $999^2 = (1000 \times 998) + 1^2 = 998,001$.) Hence the product of two 3-digit numbers must have five or six digits.

The answer to question (b) is eight or nine digits. Why? The smallest four-digit number is 1000, also known as 10^3 (a 1 followed by three zeros). The smallest five-digit number is $10,000 = 10^4$. So the smallest product is $10^3 \times 10^4 = 10^7$, which has eight digits. (Where does 10^7 comes from? $10^3 \times 10^4 = (10 \times 10 \times 10) \times (10 \times 10 \times 10 \times 10) = 10^7$.) And the largest product will be just a hair less than the ten-digit number $10^4 \times 10^5 = 10^9$, so the answer has at most nine digits.

By applying this logic, we arrive at a simple rule: **An *m*-digit number times an *n*-digit number has $m + n$ or $m + n - 1$ digits.**

It's usually easy to determine how many digits will be in the answer just by looking at the leading (leftmost) digits of each number. If the product of the leading digits is 10 or larger, then the product is guaranteed to have $m + n$ digits. (For example, with 271×828, the product of the leading digits is $2 \times 8 = 16$, so the answer has six digits.) If the product of the leading digits is 4 or smaller, then it will have $m + n - 1$ digits. For example, 314×159 has five digits. If the product of the leading digits is $5, 6, 7, 8$, or 9, then closer inspection is required. For example, 222×444 has five digits, but 234×456 has six digits. Both answers are very close to 100,000, which is really what matters.)

By reversing this rule, we get an even simpler rule for division: **An m-digit number divided by an n-digit number has $m - n$ or $m - n + 1$ digits.**

For example, a nine-digit number divided by a five-digit number must have four or five digits. The rule for determining which answer to choose is even easier than the multiplication situation. Instead of multiplying or dividing the leading digits, we simply *compare* them. If the leading digit of the first number (the number being divided) is smaller than the leading digit of the second number, then it's the smaller choice $(m - n)$. If the leading digit of the first number is larger than the leading digit of the second number, then it's the larger choice $(m - n + 1)$. If the leading digits are the same, then we look at the second digits and apply the same rule. For instance 314,159,265 divided by 12,358 will have a five-digit answer, but if we instead divide it by 62,831, the answer will have four digits. Dividing 161,803,398 by 14,142 will result in a five-digit answer since 16 is greater than 14.

I won't go through the process of doing mental division since it is similar to the pencil-and-paper method. (Indeed, any method for division problems on paper requires you to generate the answer from left to right!) But here are some shortcuts that can sometimes be handy.

When dividing by 5 (or any number ending in 5), the problem usually simplifies if you double the numerator and denominator. For example:

$$34 \div 5 = 68 \div 10 = 6.8$$

$$123 \div 4.5 = 246 \div 9 = 82 \div 3 = 27\frac{1}{3}$$

After doubling both numbers, you might notice that both 246 and 9 are divisible by 3 (we'll say more about this in Chapter 3), so we can simplify the division problem further by dividing both numbers by 3.

✂ Aside

Look at the *reciprocals* of all the numbers from 1 to 10:

$$1/2 = 0.5, \ 1/3 = 0.333\ldots, \ 1/4 = 0.25, \ 1/5 = 0.2$$

$$1/6 = 0.1666\ldots, \ 1/8 = 0.125, \ 1/9 = 0.111\ldots, \ 1/10 = 0.1$$

All of the decimal expansions above either terminate or repeat after two terms. But the one weird exception is the fraction for $1/7$, which repeats after six decimal places:

$$1/7 = 0.142857 \ 142857\ldots$$

(The reason all the other reciprocals end so quickly is that the other numbers from 2 through 11 divide into either 10, 100, 1000, 9, 90, or 99, but the first nice number that 7 divides into is 999,999.) If you write the decimal digits of $1/7$ in a circle, something magical happens:

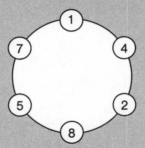

The 7th circle

What's remarkable is that all the other fractions with denominator $1/7$ can also be created by going around the circle forever from the appropriate starting point. Specifically,

$$1/7 = 0.142857 \ 142857\ldots, \quad 2/7 = 0.285714 \ 285714\ldots,$$

$$3/7 = 0.428571 \ 428571\ldots, \quad 4/7 = 0.571428 \ 571428\ldots,$$

$$5/7 = 0.714285 \ 714285\ldots, \quad 6/7 = 0.857142 \ 857142\ldots$$

Let's end this chapter with the question we asked a few pages ago. *What is the sum of all the numbers in the multiplication table?* When you first read the question, it seems intimidating, just as summing the first 100 numbers might have. By becoming more familiar with the beautiful patterns that emerge when numbers dance, we have a better chance of finding a beautiful answer to this question.

We begin by adding the numbers in the first row. As Gauss (or our triangular number formula or just simple addition) could tell us:

$$1 + 2 + 3 + 4 + 5 + 6 + 7 + 8 + 9 + 10 = 55$$

What about the sum of the second row? Well, that's just

$$2 + 4 + 6 + \cdots + 20 = 2(1 + 2 + 3 + \cdots + 10) = 2 \times 55$$

By the same reasoning, the third row will sum to 3×55. Continuing this logic, we conclude that the sum of all of these numbers is

$$(1 + 2 + 3 + \cdots + 10) \times 55 = 55 \times 55 = 55^2$$

which you should now be able to do in your head ... 3025!

CHAPTER TWO

$$\frac{2n + 4}{2} - n = 2$$

The Magic of Algebra

Magical Introduction

My first encounter with algebra was a lesson from my father when I was a kid. He said, "Son, doing algebra is just like arithmetic, except you substitute letters for numbers. For example, $2x + 3x = 5x$ and $3y + 6y = 9y$. You got it?" I said, "I think so." He said, "Okay, then what is $5Q + 5Q$?" I confidently said, "10Q." He said, "I couldn't hear you. Can you say it louder?" So I shouted, "TEN Q!" and he said, "You're welcome!" (My dad was always much more interested in puns, jokes, and stories than teaching math, so I should have been suspicious from the start!)

My second experience with algebra was trying to understand the following magic trick.

Step 1. Think of a number from 1 to 10 (though it can be larger if you wish).

Step 2. Double that number.

Step 3. Now add 10.

Step 4. Now divide by 2.

Step 5. Now subtract the number that you started with.

I believe you are now thinking of the number 5. Right?

So what is the secret behind the magic? Algebra. Let's go through the trick step by step, beginning with Step 1. I don't know what number you started with, so let's represent it by the letter N. When we use a letter to represent an unknown number, the letter is called a *variable*.

In Step 2, you are doubling the number, so you are now thinking of $2N$. (We typically avoid using the multiplication symbol, especially since the letter x is frequently used as a variable.) After Step 3, your number is $2N + 10$. For Step 4, we divide that quantity by 2, giving us $N + 5$. Finally, we subtract the original number, which was N. After subtracting N from $N + 5$, you are left with 5. We can summarize our trick in the following table.

Step 1:	N
Step 2:	$2N$
Step 3:	$2N + 10$
Step 4:	$N + 5$
Step 5:	$N + 5 - N$
Answer:	5

Rules of Algebra

Let's start with a riddle. Find a number such that when you add 5 to it, the number triples.

To solve this riddle, let's call the unknown number x. Adding 5 to it produces $x + 5$. Tripling the original number gives us $3x$. We want those numbers to be equal, so we have to solve the equation

$$3x = x + 5$$

If we subtract x from both sides of the equation, we get

$$2x = 5$$

(Where does $2x$ come from? $3x - x$ is the same as $3x - 1x$, which equals $2x$.) Dividing both sides of that equation by 2 gives us

$$x = 5/2 = 2.5$$

We can verify that this answer works, since $2.5 + 5 = 7.5$, which is the same as 3 times 2.5.

⋈ Aside

Here's another trick that algebra can help us explain. Write down any three-digit number where the digits are in decreasing order, like 842 or 951. Then reverse those numbers and subtract the second number from the first. Whatever your answer is, reverse it, then add those two numbers together. Let's illustrate with the number 853.

$$
\begin{array}{r}
853 \\
-\ 358 \\
\hline
495
\end{array}
\qquad
\begin{array}{r}
495 \\
+\ 594 \\
\hline
1089
\end{array}
$$

Now try it with a different number. What did you get? Remarkably, as long as you follow the instructions properly, you will always end up with 1089! Why is that?

Algebra to the rescue! Suppose we start with the three-digit number abc where $a > b > c$. Just as the number $853 = (8 \times 100) + (5 \times 10) + 3$, the number abc has value $100a + 10b + c$. When we reverse the digits, we get cba, which has value $100c + 10b + a$. Subtracting, we get

$$
\begin{aligned}
(100a + 10b + c) &- (100c + 10b + a) \\
&= (100a - a) + (10b - 10b) + (c - 100c) \\
&= 99a - 99c = 99(a - c)
\end{aligned}
$$

In other words, the difference has to be a multiple of 99. Since the original number has digits in decreasing order, $a - c$ is at least 2, so it must be $2, 3, 4, 5, 6, 7, 8,$ or 9. Consequently, after subtracting, we are guaranteed to have one of these numbers:

$$198, 297, 396, 495, 594, 693, 792, \text{ or } 891$$

In each of these situations, when we add the number to its reversal,

$$198 + 891 = 297 + 792 = 396 + 693 = 495 + 594 = 1089$$

we see that we are forced to end up with 1089.

We have just illustrated what I call **the golden rule of algebra: do unto one side as you would do unto the other.**

For example, suppose you wish to solve for x in the equation

$$3(2x + 10) = 90$$

Our goal is to isolate x. Let's begin by dividing both sides by 3, so the equation simplifies to

$$2x + 10 = 30$$

Next, let's get rid of that 10 by subtracting 10 from both sides. When we do that, we get

$$2x = 20$$

Finally, when we divide both sides by 2, we are simply left with

$$x = 10$$

It's always a good idea to check your answer. Here, we see that when $x = 10$, $3(2x + 10) = 3(30) = 90$, as desired. Are there any other solutions to the original equation? No, because such a value of x would also have to satisfy the subsequent equations, so $x = 10$ is the only solution.

Here's a real-life algebra problem that comes from the *New York Times*, which reported in 2014 that the movie *The Interview*, produced by Sony Pictures, generated $15 million in online sales and rentals during its first four days of availability. Sony did not say how much of this total came from $15 online sales versus $6 online rentals, but the studio did say that there were about 2 million transactions overall. To solve the reporter's problem, let's let S denote the number of online sales and let R denote the number of online rentals. Since there were 2 million transactions, we know that

$$S + R = 2{,}000{,}000$$

and since each online sale is worth $15 and each online rental is worth $6, then the total revenue satisfies

$$15S + 6R = 15{,}000{,}000$$

From the first equation, we see that $R = 2{,}000{,}000 - S$. This allows us to rewrite the second equation as

$$15S + 6(2{,}000{,}000 - S) = 15{,}000{,}000$$

or equivalently, $15S + 12{,}000{,}000 - 6S = 15{,}000{,}000$, which only uses the variable S. This can be rewritten as

$$9S + 12{,}000{,}000 = 15{,}000{,}000$$

Subtracting 12,000,000 from both sides gives us

$$9S = 3{,}000{,}000$$

and therefore S is approximately one-third of a million. That is $S \approx$ 333,333 and so $R = 2{,}000{,}000 - S \approx 1{,}666{,}667$. (Checking: total sales would be $\$15(333{,}333) + \$6(1{,}666{,}667) \approx \$15{,}000{,}000$.)

It's time to discuss a rule that we have been using throughout this book without explicitly naming it, called **the distributive law**, which is the rule that allows multiplication and addition to work well together. The distributive law says that for any numbers a, b, c,

$$a(b + c) = ab + ac$$

This is the rule that we are using when we multiply a one-digit number by a two-digit number. For example,

$$7 \times 28 = 7 \times (20 + 8) = (7 \times 20) + (7 \times 8) = 140 + 56 = 196$$

This makes sense if we think about counting. Suppose I have 7 bags of coins, and each bag has 20 gold coins and 8 silver coins. How many coins are there altogether? On the one hand, each bag has 28 coins, so the total number of coins is 7×28. On the other hand, we can also see that there are 7×20 gold coins and 7×8 silver coins, and therefore $(7 \times 20) + (7 \times 8)$ coins altogether. Consequently, $7 \times 28 = (7 \times 20) + (7 \times 8)$.

You can also view the distributive law geometrically by looking at the area of a rectangle from two different perspectives, as in the picture below.

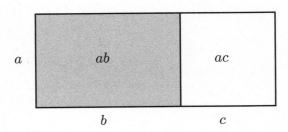

The rectangle illustrates the distributive law: $a(b + c) = ab + ac$

On the one hand, the area is $a(b + c)$. But the left part of the rectangle has area ab and the right part has area ac, so the combined area is

$ab + ac$. This illustrates the distributive law whenever a, b, c are positive numbers.

By the way, we sometimes apply the distributive law to numbers and variables together. For instance,

$$3(2x + 7) = 6x + 21$$

When this equation is read from left to right, it can be interpreted as a way of multiplying 3 times $2x + 7$. When the equation is read from right to left, it can be seen as a way of factoring $6x + 21$ by "pulling out a 3" from $6x$ and 21.

✗ **Aside**

Why does a negative number times a negative number equal a positive number? For instance, why should $(-5) \times (-7) = 35$? Teachers come up with many ways to explain this, from talking about canceling debts to simply saying "that's just the way it is." But the *real* reason is that we want the distributive law to work for *all* numbers, not just for positive numbers. And if you want the distributive law to work for negative numbers (and zero), then you must accept the consequences. Let's see why.

Suppose you accept the fact that $-5 \times 0 = 0$ and $-5 \times 7 = -35$. (These can be proved too, using a strategy similar to what we are about to do, but most people are happy to accept these statements as true.) Now evaluate the expression

$$-5 \times (-7 + 7)$$

What does this equal? On the one hand, this is just -5×0, which we know to be 0. On the other hand, using the distributive law, it must also be $((-5) \times (-7)) + (-5 \times 7)$. Consequently,

$$((-5) \times (-7)) + (-5 \times 7) = ((-5) \times (-7)) - 35 = 0$$

And since $((-5) \times (-7)) - 35 = 0$, we are forced to conclude that $(-5) \times (-7) = 35$. In general, the distributive law ensures that $(-a) \times (-b) = ab$ for all numbers a and b.

The Magic of FOIL

One important consequence of the distributive law is the **FOIL rule of algebra,** which says that for any numbers or variables $a, b, c, d,$

$$(a + b)(c + d) = ac + ad + bc + bd$$

FOIL gets its name from **First-Outer-Inner-Last**. Here, ac is the product of the *first* terms in $(a + b)(c + d)$. Then ad is the product of the *outer* terms. Then bc is the product of the *inner* terms. Then bd is the product of the *last* terms.

To illustrate, let's multiply two numbers using FOIL:

$$23 \times 45 = (20 + 3)(40 + 5)$$
$$= (20 \times 40) + (20 \times 5) + (3 \times 40) + (3 \times 5)$$
$$= 800 + 100 + 120 + 15$$
$$= 1035$$

✕ Aside

Why does FOIL work? By the distributive law (with the sum part written first) we have

$$(a + b)e = ae + be$$

Now if we replace e with $c + d$, this gives us

$$(a + b)(c + d) = a(c + d) + b(c + d) = ac + ad + bc + bd$$

where the last equality comes from applying the distributive law again. Or if you prefer a more geometric argument (when a, b, c, d are positive), find the area of the rectangle below in two different ways.

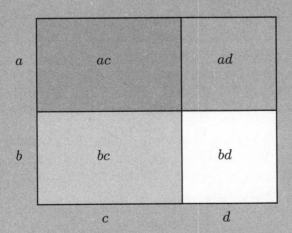

On the one hand, the rectangle has area $(a + b)(c + d)$. On the other hand, we can decompose the big rectangle into four smaller rectangles, with areas $ac, ad, bc,$ and bd. Hence the area is also equal to $ac + ad + bc + bd$. Equating these two areas gives us FOIL.

Here is a magical application of FOIL. Roll two dice and follow the instructions in the table below. As an example, let's suppose when you roll the dice, the first die has 6 on top and the second die has 3 on top. Their bottom numbers are 1 and 4, respectively.

Roll two dice (say we get 6 and 3):			
Multiply top numbers:	6×3	$=$	18
Multiply bottom numbers:	1×4	$=$	4
Multiply first top by second bottom number:	6×4	$=$	24
Multiply first bottom by second top number:	1×3	$=$	3
Total:			49

In our example, we arrived at a total of 49. And if you try it yourself with any normal six-sided dice you will arrive at the same total. It's based on the fact that on every normal six-sided die, the opposite sides add to 7. So if the dice show x and y on the top, then they must have $7 - x$ and $7 - y$ on the bottom. So using algebra, our table looks like this.

Roll 2 dice (x and y):			
Multiply top numbers:	xy	$=$	xy
Multiply bottom numbers:	$(7-x)(7-y)$	$=$	$49 - 7y - 7x + xy$
First top times second bottom:	$x(7-y)$	$=$	$7x - xy$
First bottom times second top:	$(7-x)y$	$=$	$7y - xy$
Total:		$=$	49

Notice how we use FOIL in the third row (and that $-x$ times $-y$ is positive xy). We could also arrive at 49 using less algebra by looking at the second column of our table and noticing that these are precisely the four terms we would get by FOILing $(x + (7 - x))(y + (7 - y)) = 7 \times 7 = 49$.

In most algebra classes, FOIL is mainly used for multiplying expressions like the ones below.

$$(x + 3)(x + 4) = x^2 + 4x + 3x + 12 = x^2 + 7x + 12$$

Notice that in our final expression, 7 (called the *coefficient* of the x term) is just the sum of the two numbers $3 + 4$. And the last number, 12 (called the *constant term*) is the product of the numbers 3×4. With practice, you can immediately write down the product. For example, since $5 + 7 = 12$ and $5 \times 7 = 35$, we instantly get

$$(x + 5)(x + 7) = x^2 + 12x + 35$$

It works with negative numbers, too. Here are some examples. In the first example, we exploit the fact that $6 + (-2) = 4$ and $6 \times (-2) = -12$.

$$(x + 6)(x - 2) = x^2 + 4x - 12$$
$$(x + 1)(x - 8) = x^2 - 7x - 8$$
$$(x - 5)(x - 7) = x^2 - 12x + 35$$

Here are examples where the numbers are the same.

$$(x + 5)^2 = (x + 5)(x + 5) = x^2 + 10x + 25$$
$$(x - 5)^2 = (x - 5)(x - 5) = x^2 - 10x + 25$$

Notice that, in particular, $(x + 5)^2 \neq x^2 + 25$, a common mistake made by beginning algebra students. On the other hand, something interesting does happen when the numbers have opposite signs. For example, since $5 + (-5) = 0$,

$$(x + 5)(x - 5) = x^2 + 5x - 5x - 25 = x^2 - 25$$

In general, it's worth remembering the *difference of squares* formula:

$$(x + y)(x - y) = x^2 - y^2$$

We applied this formula in Chapter 1, where we learned a shortcut for squaring numbers quickly. The method was based on the following algebra:

$$A^2 = (A + d)(A - d) + d^2$$

Let's first verify the formula. By the difference of squares formula, we see that $[(A + d)(A - d)] + d^2 = [A^2 - d^2] + d^2 = A^2$. Hence the formula works for all values of A and d. In practice, A is the number that is being squared, and d is the distance to the nearest easy number. For example, to square 97, we choose $d = 3$ so that

$$97^2 = (97 + 3)(97 - 3) + 3^2$$
$$= (100 \times 94) + 9$$
$$= 9409$$

> **⋈ Aside**
>
> Here's a proof of the difference of square law using pictures. It shows how a geometric object with area $x^2 - y^2$ can be cut and rearranged to form a rectangle with area $(x+y)(x-y)$.
>
>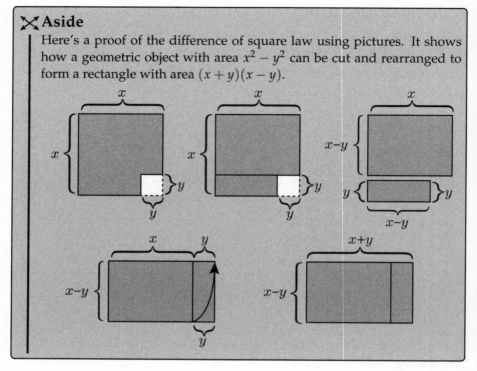

We also learned in Chapter 1 a method for multiplying numbers that were close together. We focused on numbers that were near 100 or began with the same digit, but once we understand the algebra behind it, we can apply it to more situations. Here is the algebra behind the close-together method.

$$(z+a)(z+b) = z(z+a+b) + ab$$

The formula works because $(z+a)(z+b) = z^2 + zb + za + ab$, and then we can factor out z from the first three terms. The formula works for any numbers, but we typically choose z to end in zero (which is why I chose the letter z). For example, to do the problem 43×48, we let $z = 40, a = 3, b = 8$. Then our formula tells us that

$$
\begin{aligned}
43 \times 48 &= (40+3)(40+8) \\
&= 40(40+3+8) + (3 \times 8) \\
&= (40 \times 51) + (3 \times 8) \\
&= 2040 + 24 \\
&= 2064
\end{aligned}
$$

Notice that the original numbers being multiplied have a sum of $43 + 48 = 91$, and the easier numbers being multiplied also have a sum of $40 + 51 = 91$. This is not a coincidence, since the algebra tells us that the original numbers being multiplied have a sum of $(z + a) + (z + b) = 2z + a + b$, and this is also the sum of the easier numbers z and $z + a + b$. With this algebra, we see that we could also round *up* to easy numbers. For instance, the last calculation could also have been performed with $z = 50$, $a = -7$, and $b = -2$, so our initial multiplication will be 50×41. (An easy way to get 41 is to notice that $43 + 48 = 91 = 50 + 41$.) Consequently,

$$43 \times 48 = (50 - 7)(50 - 2)$$
$$= (50 \times 41) + (-7 \times -2)$$
$$= 2050 + 14$$
$$= 2064$$

⋊Aside

In Chapter 1, we used this method for multiplying numbers that were just above 100, but it also works magically with numbers that are just below. For example,

$$96 \times 97 = (100 - 4)(100 - 3)$$
$$= (100 \times 93) + (-4 \times -3)$$
$$= 9300 + 12$$
$$= 9312$$

Note that $96 + 97 = 193 = 100 + 93$. (In practice, I just add the last digits, $6 + 7$, to know that 100 will be multiplied by a number that ends in 3, so it must be 93.) Also, once you get the hang of it, you don't have to multiply two negative numbers together, but just multiply their positive values. For example,

$$97 \times 87 = (100 - 3)(100 - 13)$$
$$= (100 \times 84) + (3 \times 13)$$
$$= 8400 + 39$$
$$= 8439$$

(continues on the following page)

⋈ Aside (*continued*)

The method can also be applied to numbers that are just below and above 100, but now you have to do a subtraction at the end. For instance,

$$109 \times 93 = (100 + 9)(100 - 7)$$
$$= (100 \times 102) - (9 \times 7)$$
$$= 10,200 - 63$$
$$= 10,137$$

Again, the number 102 can be obtained through $109 - 7$ or $93 + 9$ or $109 + 93 - 100$ (or just by summing the last digits of the original numbers; $9 + 3$ tells you that the number will end in 2, which may be enough information). With practice, you can use this to multiply any numbers that are relatively close together. I'll illustrate using three-digit numbers of moderate difficulty. Notice here, the numbers a and b are not one-digit numbers.

$$218 \times 211 = (200 + 18)(200 + 11)$$
$$= (200 \times 229) + (18 \times 11)$$
$$= 45,800 + 198$$
$$= 45,998$$

$$985 \times 978 = (1000 - 15)(1000 - 22)$$
$$= (1000 \times 963) + (15 \times 22)$$
$$= 963,000 + 330$$
$$= 963,330$$

Solving for x

Earlier in this chapter, we saw examples of solving some equations by applying the golden rule of algebra. When the equation contains just one variable (say x) and when both sides of the equation are *linear* (which means that they can contain numbers or multiples of x, but nothing more complicated, like x^2 terms), then it is easy to solve for x.

For example, to solve the equation

$$9x - 7 = 47$$

we could add 7 to both sides of the equation to get $9x = 54$, then divide by 9 to find $x = 6$.

Or for a slightly more complicated algebra problem:

$$5x + 11 = 2x + 18$$

we simplify by subtracting $2x$ from both sides and (at the same time, if you like) subtracting 11 from both sides, resulting in

$$3x = 7$$

which has the solution $x = 7/3$. Ultimately any linear equation can be simplified to $ax = b$ (or $ax - b = 0$) which has solution $x = b/a$ (assuming $a \neq 0$).

The situation gets more complicated for *quadratic* equations (where the variable x^2 enters the picture). The easiest quadratic equations to solve are those like

$$x^2 = 9$$

which has *two* solutions, $x = 3$ and $x = -3$. Even when the right side is not a perfect square, as below,

$$x^2 = 10$$

we have two solutions, $x = \sqrt{10} = 3.16\ldots$ and $x = -\sqrt{10} = -3.16\ldots$. In general, for $n > 0$, the number $\sqrt{n}$, called the *square root* of n, denotes the positive number with a square of n. When n is not a perfect square, $\sqrt{n}$ is usually computed with a calculator.

✕ Aside

What about the equation $x^2 = -9$? For now, we say it has no solution. And indeed there are no *real numbers* with a square of -9. But in Chapter 10, we will see that in a very real sense, there are two solutions to this equation, namely $x = 3i$ and $x = -3i$, where i is called an *imaginary* number with a square of -1. If that sounds impossible and ridiculous now, that's fine. But there was a time in your life when *negative* numbers seemed impossible too. (How can a number be less than 0?) You just needed to look at numbers in the right (or left!) way before they made sense.

An equation like

$$x^2 + 4x = 12$$

is a little more complicated to solve because of the $4x$ term, but there are a few different ways to go about it. Just like with mental mathematics, there is often more than one way to solve the problem.

The first method that I try on problems like this is called the *factoring method*. The first step is to move everything to the left side of the equation, so all that remains on the right side is 0. Here, the equation becomes

$$x^2 + 4x - 12 = 0$$

Now what? Well, as luck would have it, in the last section, when practicing our FOILing, we saw that $x^2 + 4x - 12 = (x + 6)(x - 2)$. Hence, our problem can be transformed to

$$(x + 6)(x - 2) = 0$$

The only way that the product of two quantities can be 0 is if at least one of those quantities is 0. Consequently, we must have either $x + 6 = 0$ or $x - 2 = 0$, which means that either

$$x = -6 \text{ or } x = 2$$

which, you should verify, solves the original problem.

According to FOIL, $(x + a)(x + b) = x^2 + (a + b)x + ab$. This makes factoring a quadratic a bit like solving a riddle. For instance, in the last problem, we had to find two numbers a and b with a sum of 4 and a product of -12. The answer, $a = 6$ and $b = -2$, gives us our factorization. For practice, try to factor $x^2 + 11x + 24$. The riddle becomes: find two numbers with sum 11 and product 24. Since the numbers 3 and 8 do the trick, we have $x^2 + 11x + 24 = (x + 3)(x + 8)$.

But now suppose we have an equation like $x^2 + 9x = -13$. There is no easy way to factor $x^2 + 9x + 13$. But have no fear! In cases like this, we are rescued by **the quadratic formula**. This useful formula says that

$$ax^2 + bx + c = 0$$

has the solution

$$x = \frac{-b \pm \sqrt{b^2 - 4ac}}{2a}$$

where the $\pm$ symbol means "plus or minus." Here's an example. For the equation

$$x^2 + 4x - 12 = 0$$

we have $a = 1$, $b = 4$, and $c = -12$.

Hence the formula tells us

$$x = \frac{-4 \pm \sqrt{16 - 4(1)(-12)}}{2} = \frac{-4 \pm \sqrt{64}}{2} = \frac{-4 \pm 8}{2} = -2 \pm 4$$

So $x = -2 + 4 = 2$ or $x = -2 - 4 = -6$, as desired. I think you'll agree that the factoring method was more straightforward for this problem.

✂ Aside

Another interesting method for solving a quadratic equation is called *completing the square*. For the equation $x^2 + 4x = 12$, let's add 4 to both sides of the equation, so that we get

$$x^2 + 4x + 4 = 16$$

The reason we added 4 to both sides was so that the left side would become $(x + 2)(x + 2)$. Thus our problem becomes

$$(x + 2)^2 = 16$$

In other words, $(x + 2)^2 = 4^2$. Thus,

$$x + 2 = 4 \text{ or } x + 2 = -4$$

which tells us that $x = 2$ or $x = -6$, as previously observed.

But for the equation

$$x^2 + 9x + 13 = 0$$

our best option is to use the quadratic formula. Here, we have $a = 1$, $b = 9$, and $c = 13$. Therefore, the formula tells us

$$x = \frac{-9 \pm \sqrt{81 - 52}}{2} = \frac{-9 \pm \sqrt{29}}{2}$$

which is not something we would have easily noticed before. There are very few formulas that you need to memorize in mathematics, but the quadratic formula is certainly one of them. With just a little practice, you'll soon find that applying this formula is as easy as . . . a, b, c!

✂ Aside

So why does the quadratic formula work? Let's rewrite the equation $ax^2 + bx + c = 0$ as

$$ax^2 + bx = -c$$

then divide both sides by a (which is not 0) to get

$$x^2 + \frac{b}{a}x = \frac{-c}{a}$$

And since $(x + \frac{b}{2a})^2 = x^2 + \frac{b}{a}x + \frac{b^2}{4a^2}$, we can *complete the square* by adding $\frac{b^2}{4a^2}$ to both sides of the above equation, giving us

$$\left(x + \frac{b}{2a}\right)^2 = \frac{b^2}{4a^2} + \frac{-c}{a} = \frac{b^2 - 4ac}{4a^2}$$

Taking the square root of both sides,

$$x + \frac{b}{2a} = \pm\frac{\sqrt{b^2 - 4ac}}{2a}$$

Thus,

$$x = \frac{-b \pm \sqrt{b^2 - 4ac}}{2a}$$

as desired.

Algebra Made Visual Through Graphs

Mathematics took a giant step forward in the seventeenth century when the French mathematicians Pierre de Fermat and René Descartes independently discovered how algebraic equations can be visualized and, conversely, how geometrical objects can be expressed through algebraic equations.

Let's start with the graph of a simple equation

$$y = 2x + 3$$

This equation says that for every value of the variable x, we double and add 3 to obtain y. Here is a table listing a handful of values of x, y pairs. Next we plot the points, as below. When drawn on a graph, the points can be labeled as *ordered pairs*. For instance, the plotted points here would be $(-3, -3), (-2, -1), (-1, 1)$, and so on. When you connect the dots and extrapolate, the resulting object is called a *graph*. Below, we show the graph of the equation $y = 2x + 3$.

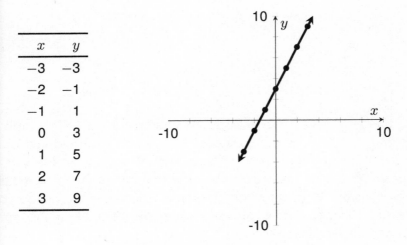

x	y
-3	-3
-2	-1
-1	1
0	3
1	5
2	7
3	9

The graph of the equation $y = 2x + 3$

Here is some useful terminology. The horizontal line in our picture is called the *x-axis*; the vertical line is the *y-axis*. The graph in this example is a *line* with *slope* 2 and *y-intercept* 3. The slope measures the steepness of the line. With a slope of 2, this says that as x increases by 1, then y increases by 2 (which you can see from the table). The *y-intercept* is simply the value of y when $x = 0$. Geometrically, this is where the line intersects the y-axis. In general, the graph of the equation

$$y = mx + b$$

is a line with slope m and y-intercept b (and vice versa). We usually identify a line with its equation. So we could simply say that the graph in the figure above is the line $y = 2x + 3$.

Here's a graph of the lines $y = 2x - 2$ and $y = -x + 7$.

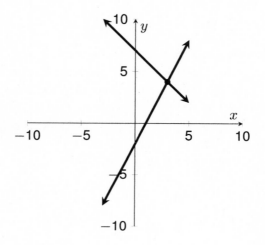

Where do the graphs of $y = 2x - 2$ and $y = -x + 7$ intersect?

The line $y = 2x - 2$ has slope 2 and y-intercept -2. (The graph is *parallel* to $y = 2x + 3$, where the entire line has been shifted vertically down by 5.) The graph $y = -x + 7$ has slope -1, so as x increases by 1, y decreases by 1. Let's use algebra to determine the point (x, y) where the lines cross. At the point where they cross, they have the same x and y value, so we want to find a value of x where the y-values are the same. In other words, we need to solve

$$2x - 2 = -x + 7$$

Adding x to both sides and adding 2 to both sides tells us that

$$3x = 9$$

so $x = 3$. Once we know x, we can use either equation to give us y. Since $y = 2x - 2$, then $y = 2(3) - 2 = 4$. (Or $y = -x + 7$ gives us $y = -3 + 7 = 4$.) Hence the lines cross at the point $(3, 4)$.

Drawing the graph of a line is easy since once you know any two points on the line, you can easily draw the entire line. The situation becomes trickier with quadratic functions (when the variable x^2 enters the picture). The simplest quadratic to graph is $y = x^2$, pictured below. Graphs of quadratic functions are called *parabolas*.

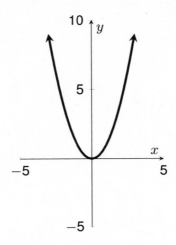

The graph of $y = x^2$

Here's the graph of $y = x^2 + 4x - 12 = (x + 6)(x - 2)$.

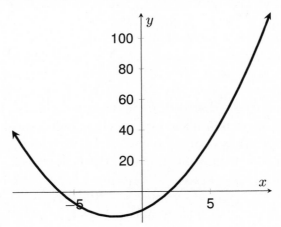

The graph of $y = x^2 + 4x - 12 = (x + 6)(x - 2)$. The y-axis has been rescaled.

Notice that when $x = -6$ or $x = 2$, then $y = 0$. We can see this in the graph since the parabola intersects the x-axis at those two points. Not coincidentally, the parabola is at its lowest point halfway between those two points when $x = -2$. The point $(-2, -16)$ is called the *vertex* of the parabola.

We encounter parabolas every day of our lives. Anytime an object is tossed, the curve created by the object is almost exactly a parabola, whether the object is a baseball or water streaming out of a fountain, as illustrated below. Properties of parabolas are also exploited in the design of headlights, telescopes, and satellite dishes.

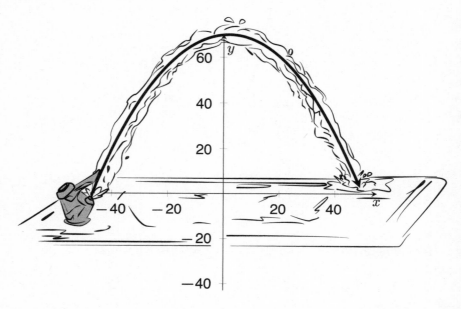

A typical water fountain. This one corresponds to the parabola $y = -.03x^2 + .08x + 70$.

Time for some terminology. So far, we have been working with *polynomials*, which are combinations of numbers and a single variable (say x), where the variable x can be raised to a positive integer power. The largest exponent is called the *degree* of the polynomial. For example, $3x + 7$ is a (linear) polynomial of degree 1. A polynomial of degree 2, like $x^2 + 4x - 12$, is called *quadratic*. A third-degree polynomial, like $5x^3 - 4x^2 - \sqrt{2}$, is called *cubic*. Polynomials of degree 4 and 5 are called *quartics* and *quintics*, respectively. (I haven't heard of names for polynomials of higher degree, mainly because they don't arise that often in practice, although I wonder if we would call seventh-degree polynomials septics. Some people might call them that, but I'm skeptical.) A polynomial with no variable in it, like the polynomial 17, has degree 0 and is called a *constant* polynomial. Finally, a polynomial is not allowed to have an infinite number of terms in it. For instance, $1 + x + x^2 + x^3 + \cdots$ is not a polynomial. (It's called an *infinite series*, which we will talk more about in Chapter 12.)

Note that with polynomials, the exponents of the variables can only be positive integers, so exponents may not be negative or fractional. For instance, if our equation includes something like $y = 1/x$ or $y = \sqrt{x}$, then it is not a polynomial since, as we'll see, $1/x = x^{-1}$ and $\sqrt{x} = x^{1/2}$.

We define the *roots* of the polynomial to be the values of x for which the polynomial equals 0. For example, $3x + 7$ has one root, namely $x = -7/3$. And the roots of $x^2 + 4x - 12$ are $x = 2$ and $x = -6$. A polynomial like $x^2 + 9$ has no (real) roots. Notice that every polynomial of degree 1 (a line) has exactly one root, since it crosses the x-axis at exactly one point, and a quadratic polynomial (a parabola) has at most two roots. The polynomials $x^2 + 1$, x^2, and $x^2 - 1$ have zero, one, and two roots, respectively.

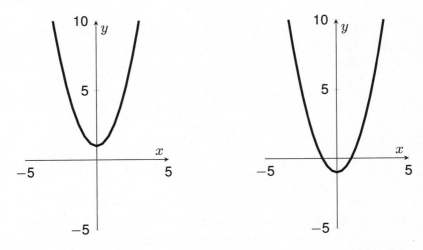

The graphs of $y = x^2 + 1$ and $y = x^2 - 1$ have, respectively, zero and two roots. The graph of $y = x^2$, pictured earlier, has just one root.

On the next page we have graphs of some cubic polynomials, and you will notice that they contain at most three roots.

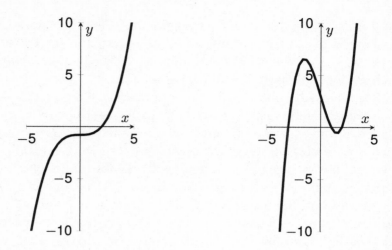

The graph of $y = (x^3 - 8)/10 = \frac{1}{10}(x-2)(x^2 + 2x + 4)$ has one root and
$y = (x^3 - 7x + 6)/2 = \frac{1}{2}(x+3)(x-1)(x-2)$ has three roots

In Chapter 10 we will encounter *the fundamental theorem of algebra*, which shows that every polynomial of degree n has at most n roots. Moreover, it can be factored into linear or quadratic parts. For example,

$$(x^3 - 7x + 6)/2 = \frac{1}{2}(x-1)(x-2)(x+3)$$

has three roots (1, 2, and -3). Whereas

$$x^3 - 8 = (x-2)(x^2 + 2x + 4)$$

has exactly one real root, when $x = 2$. (It also has two *complex* roots, but that will definitely have to wait for Chapter 10.) By the way, I should point out that nowadays it is easy to find the graph of most functions, simply by typing the equation into your favorite search engine. For instance, typing something like "y = (x^3 - 7x + 6)/2" produces a graph like the one above.

In this chapter, we have seen how to easily find the roots of any linear or quadratic polynomial. As it turns out, there are also formulas for finding the roots of a cubic or quartic polynomial, but they are extremely complicated. These formulas were discovered in the sixteenth century, and for more than two hundred years, mathematicians searched for a formula that would solve any quintic polynomial. This problem was attempted by many of the best minds in mathematics, all without success, until the Norwegian mathematician Niels Abel proved, in the early nineteenth century, that such a formula would be impossible for polynomials of degree 5 or higher. This leads to a riddle that only mathematicians find funny: Why didn't Isaac Newton prove the impossibility theorem for quintics? He wasn't Abel! We'll see examples of how to prove things impossible in Chapter 6.

> ## ✖ Aside
>
> Why does $x^{-1} = 1/x$? For example, why should $5^{-1} = 1/5$? Look at the pattern of numbers below:
>
> $$5^3 = 125, \ 5^2 = 25, \ 5^1 = 5, \ 5^0 = ?, \ 5^{-1} = ??, \ 5^{-2} = ???$$
>
> Notice that each time our exponent decreases by 1, the number is divided by 5, which makes sense if you think about it. For that pattern to continue, we would need $5^0 = 1$ and $5^{-1} = 1/5, 5^{-2} = 1/25$, and so on. But the *real* reason for it is because of the **law of exponents,** which says $x^a x^b = x^{a+b}$. Now the law makes perfect sense when a and b are positive integers. For instance, $x^2 = x \cdot x$ and $x^3 = x \cdot x \cdot x$. Thus,
>
> $$x^2 \cdot x^3 = (x \cdot x) \cdot (x \cdot x \cdot x) = x^5$$
>
> Since we want the law to work for 0, then that would require
>
> $$x^{a+0} = x^a \cdot x^0$$
>
> and since the left side equals x^a, so must the right side. But that can only happen if $x^0 = 1$.
>
> Since we want the law of exponents to work for negative integers too, then this forces us to accept
>
> $$x^1 \cdot x^{-1} = x^{1+(-1)} = x^0 = 1$$
>
> Dividing both sides by x implies that x^{-1} must equal $1/x$. By a similar argument, $x^{-2} = 1/x^2, x^{-3} = 1/x^3$, and so on.
>
> And since we want the law of exponents to be true for all real numbers as well, that forces us to accept
>
> $$x^{1/2} x^{1/2} = x^{1/2 \, + \, 1/2} = x^1 = x$$
>
> Hence when we multiply $x^{1/2}$ by itself, we get x, and therefore (when x is a positive number) we have $x^{1/2} = \sqrt{x}$.

Figuring Out Y (and X Too!)

Let's end this chapter as we began it, with an algebra-based magic trick.

Step 1. Think of two numbers from 1 to 10.

Step 2. Add those numbers together.

Step 3. Multiply that number by 10.

Step 4. Now add the larger original number.

Step 5. Now subtract the smaller original number.

Step 6. Tell me the number you're thinking of and I'll tell you both of the original numbers!

Believe it or not, with that one piece of information, you can determine *both* of the original numbers. For example, if the final answer is 126, then you must have started with 9 and 3. Even if this trick is repeated a few times, it's hard for your audience to figure out how you are doing it.

Here's the secret. To determine the larger number, take the last digit of the answer (6 here) and add it to the preceding number(s) (12 here), then divide by 2. Here we conclude that the larger digit is $(12+6)/2 = 18/2 = 9$. For the smaller digit, take the larger digit that you just computed (9) and subtract the last digit of the answer. Here that would be $9 - 6 = 3$.

Here are two more examples for practice. If the answer is 82, then the larger number is $(8+2)/2 = 5$ and the smaller number is $5 - 2 = 3$. If the answer is 137, then the larger number is $(13+7)/2 = 10$ and the smaller number is $10 - 7 = 3$.

Why does it work? Suppose you start with two numbers X and Y, where X is equal to or larger than Y. Following the original instructions and the algebra in the table below, we see that after Step 5, you end up with the number $10(X+Y) + (X-Y)$.

Step 1:	X and Y
Step 2:	$X + Y$
Step 3:	$10(X+Y)$
Step 4:	$10(X+Y) + X$
Step 5:	$10(X+Y) + (X-Y)$
Larger Number:	$((X+Y) + (X-Y))/2 = X$
Smaller Number:	$X - (X-Y) = Y$

How does that help us? Notice that a number of the form $10(X + Y)$ would have to end in 0, and the digit (or digits) preceding the 0 would be $X + Y$. Since X and Y are between 1 and 10, and X is greater than or equal to Y, then $X - Y$ is forced to be a one-digit number (between 0 and 9). Hence the last digit of the answer must be $X - Y$. For example, if you started with digits 9 and 3, then $X = 9$ and $Y = 3$. Hence the answer after Step 5 must begin with $X + Y = 9 + 3 = 12$ and end with $X - Y = 9 - 3 = 6$, resulting in 126. Once we know $X + Y$ and $X - Y$, we can take their average to get $((X + Y) + (X - Y))/2 = X$. For Y, we could calculate $((X + Y) - (X - Y))/2$ (here that would be $(12 - 6)/2 = 6/2 = 3$), but I find it simpler to just take the larger number you just calculated and then subtract the last digit of their answer ($9 - 6 = 3$) since $X - (X - Y) = Y$.

> ## ✕ Aside
>
> If you would like an additional challenge for yourself (and the spectator, who might want to use a calculator), you can ask the spectator to pick any two numbers between 1 and 100, but now in Step 3 you have them multiply their total by 100 instead of 10, then proceed as before. For example, if they started with 42 and 17, then after five steps they end up with an answer of 5925. You can reconstruct the answer by stripping off the last *two* digits from the rest of the answer and calculating their average. Here, the larger number is $(59 + 25)/2 = 84/2 = 42$. To get the smaller number, subtract the last two digits of their answer from the larger number. Here, $42 - 25 = 17$, as desired. The reason this works is almost the same explanation as before, except after step 5, the answer is $100(X + Y) - (X - Y)$, where $X - Y$ is the last two digits of the answer.
>
> Here's one more example: if the answer is 15,222 (so $X + Y = 152$ and $X - Y = 22$), then the larger number is $(152 + 22)/2 = 174/2 = 87$ and the smaller number is $87 - 22 = 65$.

CHAPTER THREE

$$\sqrt{9} = 3$$

The Magic of 9

The Most Magical Number

When I was a kid, my favorite number was 9 since it seemed to possess so many magical properties. To see an example, please follow the mathemagical instructions below.

1. Think of a number from 1 to 10 (or choose a larger whole number and use a calculator, if you'd like).

2. Triple your number.

3. Add 6 to it.

4. Now triple your number again.

5. If you wish, double your number again.

6. Add the digits of your number. If the result is a one-digit number, then stop.

7. If the sum is a two-digit number, then add the two digits together.

8. Concentrate on your answer.

I get the distinct impression that you are now thinking of the number 9. Is that right? (If not, then you may want to recheck your arithmetic.)

What is so magical about the number 9? In the rest of this chapter, we'll see some of its magical properties, and we'll even consider a world where it makes sense to say that 12 and 3 are functionally the same! The first magical property of the number 9 can be seen by looking at its multiples:

$$9, 18, 27, 36, 45, 54, 63, 72, 81, 90, 99, 108, 117, 126, 135, 144, \ldots$$

What do these numbers have in common? If you add the digits of each number, it seems that you always get 9. Let's check a few of them: 18 has digit sum $1 + 8 = 9$; 27 has $2 + 7 = 9$; 144 has $1 + 4 + 4 = 9$. But wait, there's an exception: 99 has digit sum 18, but 18 is itself a multiple of 9. So here is the key result, which you may have been taught in primary school, and which we will explain later in this chapter:

If a number is a multiple of 9, then its digit sum is a multiple of 9 (and vice versa).

For example, the number 123,456,789 has digit sum 45 (a multiple of 9), so it is a multiple of 9. Conversely, 314,159 has digit sum 23 (not a multiple of 9), so it is not a multiple of 9.

To use this rule to understand our earlier magic trick, let's examine the algebra. You started by thinking of a number, which we call N. After tripling it, you get $3N$, which becomes $3N + 6$ at the next step. Tripling that result gives us $3(3N + 6) = 9N + 18$, which equals $9(N + 2)$. If you decide to double that number, you have $18N + 36 = 9(2N + 4)$. Either way, your final answer is 9 times a whole number, and therefore you must end up with a multiple of 9. When you take the sum of the digits of this number, you must again have a multiple of 9 (probably 9 or 18 or 27 or 36), and the sum of *these* digits must be 9.

Here is a variation of the previous magic trick that I also like to perform. Ask someone with a calculator to secretly choose one of these four-digit numbers:

$$3141 \text{ or } 2718 \text{ or } 2358 \text{ or } 9999$$

These numbers are, respectively, the first four digits of π (Chapter 8), the first four digits of e (Chapter 10), consecutive Fibonacci numbers (Chapter 5), and the largest four-digit number. Then ask them to take their four-digit number and multiply it by *any* three-digit number. The end result is a six-digit or seven-digit number that you couldn't possibly know. Next ask them to mentally circle one of the digits of their answer, but not to circle a 0 (since it is already shaped like a circle!) Ask them to recite all of the uncircled digits in *any* order they want and to concentrate on the remaining digit. With just a little concentration on your part, you successfully reveal the answer.

So what's the secret? Notice that all of the four numbers they could start with are multiples of 9. Since you start with a multiple of 9, and you are multiplying it by a whole number, the answer will still be a multiple of 9. Thus, its digits must add up to a multiple of 9. As the numbers are called out to you, simply add them up. The missing digit is the number you need to add to your total to reach a multiple of 9. For example, suppose the spectator calls out the digits 5, 0, 2, 2, 6, and 1. The sum of these numbers is 16, so they must have left out the number 2 to reach the nearest multiple of 9, namely 18. If the numbers called out are 1, 1, 2, 3, 5, 8, with a total of 20, then the missing digit must be 7 to reach 27. Suppose the numbers called out to you add up to 18; what did they leave out? Since we told them not to concentrate on 0, then the missing digit must be 9.

So why do multiples of 9 always add up to multiples of 9? Let's look at an example. The number 3456, when expressed in terms of powers of 10, looks like

$$3456 = (3 \times 1000) + (4 \times 100) + (5 \times 10) + 6$$
$$= 3(999 + 1) + 4(99 + 1) + 5(9 + 1) + 6$$
$$= 3(999) + 4(99) + 5(9) + 3 + 4 + 5 + 6$$
$$= (\text{a multiple of } 9) + 18$$
$$= \text{a multiple of } 9$$

By the same logic, any number with a digit sum that is a multiple of 9 must itself be a multiple of 9 (and vice versa: any multiple of 9 must have a digit sum that is a multiple of 9).

Casting Out Nines

What happens when the sum of the digits of your number is *not* a multiple of 9? For example, consider the number 3457. Following the above process, we can write 3457 (whose digits sum to 19) as $3(999) + 4(99) + 5(9) + 7 + 12$, so 3457 is $7 + 12 = 19$ bigger than a multiple of 9. And since $19 = 18 + 1$, this indicates that 3457 is just 1 bigger than a multiple of 9. We can reach the same conclusion by adding the digits of 19, then adding the digits of 10, which I represent as

$$3457 \to 19 \to 10 \to 1$$

The process of adding the digits of your number and repeating that process until you are reduced to a one-digit number is called *casting out nines*, since at each step of the process, you are subtracting a multiple of 9. The one-digit number obtained at the end of the process is called the *digital root* of the original number. For example, the digital root of 3457 is 1. The number 3456 has digital root 9. We can succinctly summarize our previous conclusions as follows. For any positive number n:

If n has a digital root of 9, then n is a multiple of 9.
Otherwise, the digital root is the remainder obtained when n is divided by 9.

Or expressed more algebraically, if n has digital root r, then

$$n = 9x + r$$

for some integer x. The process of casting out nines can be a fun way to check your answers to addition, subtraction, and multiplication problems. For example, if an addition problem is solved correctly, then the digital root of the answer must agree with the sum of the digital roots. Here's an example. Perform the addition problem

$$
\begin{array}{r}
91787 \to 32 \to 5 \\
+\ 42864 \to 24 \to 6 \\
\hline
134651 \qquad\quad 11 \to ② \\
\downarrow \qquad\qquad\qquad \\
20 \quad \to ② \qquad
\end{array}
$$

Notice that the numbers being added have digital roots of 5 and 6 and their sum, 11, has digital root 2. It is no coincidence that the digital root of the answer, 134,651, also has a digital root of 2. The reason this process works in general is based on the algebra

$$(9x + r_1) + (9y + r_2) = 9(x + y) + (r_1 + r_2)$$

If the numbers didn't match, then you have definitely made a mistake somewhere. Important: if the numbers *do* match, then your answer is not necessarily right. But this process will catch most random errors about 90 percent of the time. Note that it will not catch errors where you've accidentally swapped two of the correct digits, since the swapping of two correct digits does not alter the digit sum. But if a single digit is in error, it will detect that mistake, unless the error turned a 0 into a 9 or a 9 into a 0. This process can be applied even when we are adding long columns of numbers. For example, suppose that you purchased a number of items with the prices below:

$$
\begin{array}{ccc}
112.56 & \rightarrow 15 \rightarrow & 6 \\
96.50 & \rightarrow 20 \rightarrow & 2 \\
14.95 & \rightarrow 19 \rightarrow & 1 \\
48.95 & \rightarrow 26 \rightarrow & 8 \\
108.00 & \rightarrow 9 \rightarrow & 9 \\
17.52 & \rightarrow 15 \rightarrow & \underline{6} \\
\overline{398.48} & & 32 \rightarrow \boxed{5} \\
\downarrow & & \\
32 \rightarrow \boxed{5} & &
\end{array}
$$

Adding the digits of our answer, we see that our total has digital root 5, and the sum of the digital roots is 32, which is consistent with our answer, since 32 has digital root 5. Casting out nines works for subtraction as well. For instance, subtract the numbers from our earlier addition problem:

$$
\begin{array}{ccc}
91787 & \rightarrow 32 \rightarrow & 5 \\
-\ 42864 & \rightarrow 24 \rightarrow & \underline{6} \\
\overline{48923} & & -1 \rightarrow \boxed{8} \\
\downarrow & & \\
26 \rightarrow \boxed{8} & &
\end{array}
$$

The answer to the subtraction problem 48,923 has digital root 8. When we subtract the digital roots of the original numbers we see that $5 - 6 = -1$. But this is consistent with our answer, since $-1 + 9 = 8$, and adding (or subtracting) multiples of 9 to our answer doesn't change the digital root. By the same logic, a difference of 0 is consistent with a digital root of 9.

Let's take advantage of what we've learned to create another magic trick (like the one given in the book's introduction). Follow these instructions; you may use a calculator if you wish.

1. Think of any two-digit or three-digit number.

2. Add your digits together.

3. Subtract that from your original number.

4. Add the digits of your new number.

5. If your total is even, then multiply it by 5.

6. If your total is odd, then multiply it by 10.

7. Now subtract 15.

Are you thinking of 75?

For example, if you started with the number 47, then you began by adding $4 + 7 = 11$, followed by $47 - 11 = 36$. Next $3 + 6 = 9$, which is an odd number. Multiplying it by 10 gives you 90, and $90 - 15 = 75$. On the other hand, if you started with a 3-digit number like 831, then $8 + 3 + 1 = 12$; $831 - 12 = 819$; $8 + 1 + 9 = 18$, an even number. Then $18 \times 5 = 90$, and subtracting 15 gives us 75, as before.

The reason this trick works is that if the original number has a digit sum of T, then the number must be T greater than a multiple of 9. When we subtract T from the original number, we are guaranteed to have a multiple of 9 below 999, so its total will be either 9 or 18. (For example, when we started with 47, it had a digit sum of 11. We subtracted 11 to get to 36, with digit sum 9.) After the next step, we will be forced to have 90 (as 9×10 or 18×5) followed by 75, as in the examples above.

Casting out nines works for multiplication too. Let's see what happens as we multiply the previous numbers:

$$
\begin{array}{r}
91787 \rightarrow 32 \rightarrow 5 \\
\times\ 42864 \rightarrow 24 \rightarrow 6 \\
\hline
3{,}934{,}357{,}968 \qquad\qquad 30 \rightarrow \textcircled{3} \\
\downarrow \\
57 \rightarrow 12 \rightarrow \textcircled{3}
\end{array}
$$

The reason that casting out nines works for multiplication is based on FOIL from Chapter 2. For instance, in our last example, the digital roots on the right tell us that the numbers being multiplied are of the form $9x + 5$ and $9y + 6$, for some integers x and y. And when we multiply these together, we get

$$
\begin{aligned}
(9x + 5)(9y + 6) &= 81xy + 54x + 45y + 30 \\
&= 9(9xy + 6x + 5y) + 30 \\
&= \text{(a multiple of 9)} + (27 + 3) \\
&= \text{(a multiple of 9)} + 3
\end{aligned}
$$

Although casting out nines is not traditionally used to check division problems, I can't resist showing you an utterly magical method for dividing numbers by 9. This method is sometimes referred to as the Vedic method. Consider the problem

$$12302 \div 9$$

Write the problem as

$$9)\overline{1\ 2\ 3\ 0\ 2}$$

Now bring the first digit above the line and write the letter R (as in remainder) above the last digit, like so.

$$\begin{array}{r} ①\qquad\ \text{R} \\ 9)\overline{1\ 2\ 3\ 0\ 2} \end{array}$$

Next we add numbers diagonally like in the circled positions below. The circled numbers 1 and 2 add to 3, so we put a 3 as the next number in the quotient.

$$\begin{array}{r} ①3\qquad \text{R} \\ 9)\overline{1②3\ 0\ 2} \end{array}$$

Then $3 + 3 = 6$.

$$\begin{array}{r} 1③6\quad \text{R} \\ 9)\overline{1\ 2③0\ 2} \end{array}$$

Then $6 + 0 = 6$.

$$\begin{array}{r} 1\ 3\ 6\ 6\ \text{R} \\ 9\overline{)1\ 2\ 3\ 0\ 2} \end{array}$$

Finally, we add $6 + 2 = 8$ for our remainder.

$$\begin{array}{r} 1\ 3\ 6\ 6\ \text{R}\ 8 \\ 9\overline{)1\ 2\ 3\ 0\ 2} \end{array}$$

And there's the answer: $12{,}302 \div 9 = 1366$ with a remainder of 8. That seemed almost too easy! Let's do another problem with fewer details.

$$31415 \div 9$$

Here's the answer!

$$\begin{array}{r} 3\ 4\ 8\ 9\ \text{R}\ 14 \\ 9\overline{)3\ 1\ 4\ 1\ 5} \end{array}$$

Starting with the 3 on top, we compute $3 + 1 = 4$, then $4 + 4 = 8$, then $8 + 1 = 9$, then $9 + 5 = 14$. So the answer is 3489 with a remainder of 14. But since $14 = 9 + 5$, we add 1 to the quotient to get an answer of 3490 with a remainder of 5.

Here's a simple question with an attractive answer. I'll leave it to you to verify (on paper or in your head) that

$$111{,}111 \div 9 = 12{,}345\ \text{R}\ 6$$

We saw that when the remainder is 9 or larger, we simply added 1 to our quotient, and subtracted 9 from the remainder. The same sort of thing happens when we have a sum that exceeds 9 in the middle of the division problem. We indicate the carry, then subtract 9 from the total and continue (or should I say *carry on*?) as before. For example, with the problem $4821 \div 9$, we start off like this:

$$\begin{array}{r} 4\ \ \ \ \ \ \text{R} \\ 9\overline{)4\ 8\ 2\ 1} \end{array}$$

Here we start with the 4, but since $4 + 8 = 12$, we place a 1 above the 4 (to indicate a carry), then subtract 9 from 12 to write 3 in the next spot. This is followed by $3 + 2 = 5$, then $5 + 1 = 6$ to get an answer of 535 with a remainder of 6, as illustrated on the next page.

$$\begin{array}{r} \overset{1}{4\,3\,5}\,R\,6 \\ 9\overline{)4\,8\,2\,1} \end{array}$$

Here is one more problem with lots of carries. Try $98{,}765 \div 9$.

$$\begin{array}{r} \overset{1\ 1\ 1}{9\,8\,6\,3}\,R\,8 \\ 9\overline{)9\,8\,7\,6\,5} \end{array}$$

Starting with 9 on top, we add $9 + 8 = 17$, indicate the carry and subtract 9, so the second digit of the quotient becomes 8. Next, $8 + 7 = 15$; indicate the carry and write $15 - 9 = 6$. Then $6 + 6 = 12$; indicate the carry and write $12 - 9 = 3$. Finally, your remainder is $3 + 5 = 8$. Taking all the carries into account, our answer is 10,973 with a remainder of 8.

> ## ⋈ Aside
>
> If you think dividing by 9 is cool, check out dividing by 91. Ask for any two-digit number and you can instantly divide that number by 91 to as many decimal places as desired. No pencils, no paper, no kidding! For example,
>
> $$53 \div 91 = 0.582417\ldots$$
>
> More specifically, the answer is $0.\overline{582417}$, where the bar above the digits 582417 means that those numbers repeat indefinitely. Where do these numbers come from? It's as easy as multiplying the original two-digit number by 11. Using the method we learned from Chapter 1, we calculate $53 \times 11 = 583$. Subtracting 1 from that number gives us the first half of our answer, namely 0.582. The second half is the first half subtracted from 999, which is $999 - 582 = 417$. Therefore, our answer is $0.\overline{582417}$, as promised.
>
> Let's do one more example. Try $78 \div 91$. Here, $78 \times 11 = 858$, so the answer begins with 857. Then $999 - 857 = 142$, so $78 \div 91 = 0.\overline{857142}$. We have actually seen that number in Chapter 1, because $78/91$ simplifies to $6/7$.
>
> This method works because $91 \times 11 = 1001$. Thus, in the first example, $\frac{53}{91} = \frac{53 \times 11}{91 \times 11} = \frac{583}{1001}$. And since $1/1001 = 0.\overline{000999}$, we get the repeating part of our answer from $583 \times 999 = 583{,}000 - 583 = 582{,}417$.
>
> Since $91 = 13 \times 7$, this gives us a nice way to divide numbers by 13 by *unsimplifying* it to have a denominator of 91. For instance, $1/13 = 7/91$, and since $7 \times 11 = 077$, we get
>
> $$1/13 = 7/91 = 0.\overline{076923}$$
>
> Likewise, $2/13 = 14/91 = 0.\overline{153846}$, since $14 \times 11 = 154$.

The Magic of 10, 11, 12, and Modular Arithmetic

Much of what we have learned about the number 9 extends to all other numbers as well. When casting out nines, we were essentially replacing numbers with their remainders when divided by 9. The idea of replacing a number with its remainder is not new for most people. We have been doing it ever since we learned how to tell time. For example, if a clock indicates that the time is 8 o'clock (without distinguishing between morning or evening), then what time will it indicate in 3 hours? What about in 15 hours? Or 27 hours? Or 9 hours ago? Although your first reaction might be to say that the hour would either be 11 or 23 or 35 or −1, as far as the clock is concerned, all those times are represented by 11 o'clock, since all of those times differ by multiples of 12 hours. The notation mathematicians use is

$$11 \equiv 23 \equiv 35 \equiv -1 \pmod{12}$$

What time will the clock say in 3 hours? In 15 hours? Or 27 hours? Or 9 hours ago?

In general, we say that $a \equiv b \pmod{12}$ if a and b differ by a multiple of 12. Equivalently, $a \equiv b \pmod{12}$ if a and b have the same remainder when divided by 12. More generally, for any positive integer m, we say that two numbers a and b are *congruent mod m*, denoted $a \equiv b \pmod{m}$, if a and b differ by a multiple of m. Equivalently,

$$a \equiv b \pmod{m} \text{ if } a = b + qm \text{ for some integer } q$$

The nice thing about congruences is that they behave almost exactly the same as regular equations, and we can perform *modular arithmetic* by adding, subtracting, and multiplying them together. For example, if

$a \equiv b \pmod{m}$ and c is any integer, then it is also true that

$$a + c \equiv b + c \text{ and } ac \equiv bc \pmod{m}$$

Different congruences can be added, subtracted, and multiplied. For instance, if $a \equiv b \pmod{m}$ and $c \equiv d \pmod{m}$, then

$$a + c \equiv b + d \text{ and } ac \equiv bd \pmod{m}$$

For example, since $14 \equiv 2$ and $17 \equiv 5 \pmod{12}$, then $14 \times 17 \equiv 2 \times 5 \pmod{12}$, and indeed $238 = 10 + (12 \times 19)$. A consequence of this rule is that we can raise congruences to powers. So if $a \equiv b \pmod{m}$, then we have the **power rule:**

$$a^2 \equiv b^2 \qquad a^3 \equiv b^3 \qquad \cdots \qquad a^n \equiv b^n \pmod{m}$$

for any positive integer n.

✕ Aside

Why does modular arithmetic work? If $a \equiv b \pmod{m}$ and $c \equiv d \pmod{m}$, then $a = b + pm$ and $c = d + qm$ for some integers p and q. Thus $a + c = (b + d) + (p + q)m$, and therefore $a + c \equiv b + d \pmod{m}$. Furthermore, by FOIL,

$$ac = (b + pm)(d + qm) = bd + (bq + pd + pqm)m$$

so ac and bd differ by a multiple of m and so $ac \equiv bd \pmod{m}$. Multiplying the congruence $a \equiv b \pmod{m}$ by itself gives us $a^2 \equiv b^2 \pmod{m}$, and repeating this process gives us the power rule.

It is the power rule that makes 9 such a special number when working in base 10. Since

$$10 \equiv 1 \pmod{9}$$

the power rule tells us that $10^n \equiv 1^n = 1 \pmod{9}$ for any n, and therefore a number like 3456 satisfies

$$
\begin{aligned}
3456 &= 3(1000) + 4(100) + 5(10) + 6 \\
&\equiv 3(1) + 4(1) + 5(1) + 6 = 3 + 4 + 5 + 6 \pmod{9}
\end{aligned}
$$

Since $10 \equiv 1 \pmod{3}$, this explains why you can also determine if a number is a multiple of 3 (or what its remainder will be when divided by 3) just by adding up its digits. If we worked in a different base, say in base 16 (called the *hexadecimal system,* used in electrical engineering and computer science), then since $16 \equiv 1 \pmod{15}$, you could tell if a number is a multiple of 15 (or 3 or 5), or determine its remainder when divided by 15, just by adding its digits.

Back to base 10. There is a neat way to determine if a number is a multiple of 11. It is based on the fact that

$$10 \equiv -1 \pmod{11}$$

and therefore $10^n \equiv (-1)^n \pmod{11}$. Therefore $10^2 \equiv 1 \pmod{11}$, $10^3 \equiv (-1) \pmod{11}$, and so on. For example, a number like 3456 satisfies

$$
\begin{aligned}
3456 &= 3(1000) + 4(100) + 5(10) + 6 \\
&\equiv -3 + 4 - 5 + 6 = 2 \pmod{11}
\end{aligned}
$$

Thus 3456 has a remainder of 2 when divided by 11. The general rule is that a number is a multiple of 11 if, and only if, we end up with a multiple of 11 (such as $0, \pm 11, \pm 22, \dots$) when we alternately subtract and add the digits. For instance, is the number 31,415 a multiple of 11? By calculating $3 - 1 + 4 - 1 + 5 = 10$, we conclude that it is not, but if we considered 31,416, then our total would be 11, and so 31,416 would be a multiple of 11.

Mod 11 arithmetic is actually used in the creation and verification of an ISBN number (International Standard Book Number). Suppose your book has a 10-digit ISBN (as most books do that were published prior to 2007). The first few digits encode the book's country of origin, publisher, and title, but the tenth digit (called the *check digit*) is chosen so that the numbers satisfy a special numerical relationship. Specifically, if the 10-digit number looks like *a-bcd-efghi-j*, then *j* is chosen to satisfy

$$10a + 9b + 8c + 7d + 6e + 5f + 4g + 3h + 2i + j \equiv 0 \pmod{11}$$

For example, my book *Secrets of Mental Math*, published in 2006, has ISBN 0-307-33840-1, and indeed

$$
\begin{aligned}
10(0) + 9(3) + 8(0) + 7(7) + 6(3) + 5(3) + 4(8) + 3(4) + 2(0) + 1 \\
= 154 \equiv 0 \pmod{11}
\end{aligned}
$$

since $154 = 11 \times 14$. You might wonder what happens if the check digit is required to be 10. In that case, the digit is assigned the letter X, which is the Roman numeral for 10. The ISBN system has the nice feature that if a single digit is entered incorrectly, the system can detect that. For instance, if the third digit is incorrect, then the total at the end will be off by some multiple of 8, either as ± 8 or ± 16 or ... ± 80. But none of those numbers are multiples of 11 (because 11 is a prime number), so the adjusted total cannot be a multiple of 11. In fact, with a little bit of algebra, it can also be shown that the system can detect an error whenever two different digits are transposed. For example, suppose that digits c and f were transposed but everything else was correct. Then the only part of the total affected is the contribution from the c and f term. The old total uses $8c + 5f$ and the new total uses $8f + 5c$. The difference is $(8f + 5c) - (8c + 5f) = 3(f - c)$, which is not a multiple of 11. Hence the new total will not be a multiple of 11.

In 2007, publishers switched to the ISBN-13 system, which used 13 digits and was based on mod 10 arithmetic instead of mod 11. Under this new system, the number $abc\text{-}d\text{-}efg\text{-}hijkl\text{-}m$ can only be valid when it satisfies

$$a + 3b + c + 3d + e + 3f + g + 3h + i + 3j + k + 3l + m \equiv 0 (\text{mod } 10)$$

For example, the ISBN-13 for this book is 978-0-465-05472-5. A quick way to check this number is to separate the odd and even positioned numbers to compute

$$(9 + 8 + 4 + 5 + 5 + 7 + 5) + 3(7 + 0 + 6 + 0 + 4 + 2)$$
$$= 43 + 3(19) = 43 + 57 = 100 \equiv 0 \quad (\text{mod } 10)$$

The ISBN-13 system will detect any single-digit error and most (but not all) transposition errors of consecutive terms. For instance, in the last example, if the last three digits are changed from 725 to 275, then the error will not be detected, since the new total will be 110, which is still a multiple of 10. A similar sort of mod 10 system is in place for verifying barcode, credit card, and debit card numbers. Modular arithmetic also plays a major role in the design of electronic circuits and Internet financial security.

Calendar Calculating

My favorite mathematical party trick is to tell people the day of the week they were born, given their birthday information. For example, if someone told you that she was born on May 2, 2002, you can instantly tell her that she was born on a Thursday. An even more practical skill is the ability to figure out the day of the week for any date of the current or upcoming year. I will teach you an easy method to do this, and the mathematics behind it, in this section.

But before we delve into the method, it's worth reviewing some of the scientific and historical background behind the calendar. Since the Earth takes about 365.25 days to travel around the Sun, a typical year has 365 days, but we add a leap day, February 29, every four years. (This way, in four years, we have $4 \times 365 + 1 = 1461$ days, which is just about right.) This was the idea behind the Julian calendar, established by Julius Caesar more than two thousand years ago. For example, the year 2000 is a leap year, as is every fourth year after that: 2004, 2008, 2012, 2016, and so on, up through 2096. Yet 2100 will not be a leap year. Why is that?

The problem is that a year is actually about 365.243 days (about eleven minutes less than 365.25), so leap years are ever so slightly over-represented. With four hundred trips around the Sun, we experience 146,097 days, but the Julian calendar allocated $400 \times 365.25 = 146{,}100$ days for this (which is three days too long). To avoid this problem (and other difficulties associated with the timing of Easter) the Gregorian calendar was established by Pope Gregory XIII in 1582. In that year, the Catholic nations removed ten days from their calendar. For example, in Spain, the Julian date of Thursday, October 4, 1582, was followed by the Gregorian date of Friday, October 15, 1582. Under the Gregorian calendar, years that were divisible by 100 would no longer be leap years, unless they were also divisible by 400 (thus removing three days). Consequently, 1600 remained a leap year on the Gregorian calendar, but 1700, 1800, and 1900 would not be leap years. Likewise, 2000 and 2400 are leap years, but the years 2100, 2200, and 2300 are not leap years. Under this system, in any four hundred year period, the number of leap years is $100 - 3 = 97$ and therefore the number of days would be $(400 \times 365) + 97 = 146{,}097$ as desired.

The Gregorian calendar was not accepted by all countries right away, and the non-Catholic countries were particularly slow to adopt it. For example, England and its colonies didn't make the switch until 1752, when Wednesday, September 2, was followed by Thursday, September 14. (Notice that eleven days were eliminated, since 1700 was a leap year on the Julian calendar but not in the Gregorian calendar.) It was not until the 1920s that all countries had converted from the Julian to the Gregorian calendar. This has been a source of complications for historians. My favorite historical paradox is that both William Shakespeare and Miguel de Cervantes died on the same date, April 23, 1616, and yet they died ten days apart. That's because when Cervantes died, Spain had converted to the Gregorian calendar, but England was still on the Julian calendar. So when Cervantes died on Gregorian April 23, 1616, it was still April 13, 1616, in England, where Shakespeare was living (if only for ten more days).

The formula to determine the day of the week for any date on the Gregorian calendar goes like this:

$$\text{Day of Week} \equiv \text{Month Code} + \text{Date} + \text{Year Code} \pmod 7$$

and we will explain all of these terms shortly. It makes sense that the formula uses modular arithmetic, working mod 7, since there are 7 days in a week. For example, if a date is 72 days in the future, then its day of the week will be two days from today, since $72 \equiv 2 \pmod 7$. Or a date that is 28 days from today will have the same day of the week, since 28 is a multiple of 7.

Let's start with the codes for the days of the week, which are easy to memorize.

Number	Day	Mnemonic
1	Monday	1-day
2	Tuesday	2s-day
3	Wednesday	raise 3 fingers
4	Thursday	4s-day
5	Friday	5-day
6	Saturday	6er-day
7 or 0	Sunday	7-day or none-day

I have provided mnemonics to go along with each number-day pair, most of which are self-explanatory. For Wednesday, notice that if you raise three fingers on your hand, you create the letter W. For Thursday, if you pronounce it as "Thor's Day," then it will rhyme with "Four's Day."

> ✂ **Aside**
>
> So where do the names of the days of the week come from? The custom of naming the days of the week after the Sun, the Moon, and the five closest heavenly bodies dates back to ancient Babylonia. From the Sun, the Moon, and Saturn, we immediately get Sunday, Monday, and Saturday. Other names are easier to see in French or Spanish. For instance, Mars becomes Mardi or Martes; Mercury becomes Mercredi or Miércoles; Jupiter becomes Jeudi or Jueves; Venus becomes Vendredi or Viernes. Note that Mars, Mercury, Jupiter, and Venus were also the names of Roman gods and goddesses. The English language has Germanic origins and the early Germans renamed some of these days for Norse gods. So Mars became Tiw, Mercury became Woden, Jupiter became Thor, and Venus became Freya, and that's how we arrived at the names for Tuesday, Wednesday, Thursday, and Friday.

The month codes are given below, along with my mnemonics for remembering them.

Month	Code	Mnemonic
January*	6	W-I-N-T-E-R
February*	2	Month number 2
March	2	March 2 the beat!
April	5	A-P-R-I-L or F-O-O-L-S
May	0	Hold the May-O!
June	3	June B-U-G
July	5	Fiver-works in the sky!
August	1	August begins with A = 1
September	4	Beginning of F-A-L-L
October	6	T-R-I-C-K-S (rhymes with 6)
November	2	2 pieces of 2rkey!
December	4	L-A-S-T or X-M-A-S

*Exception: In leap years, January = 5 and February = 1

I'll explain how these numbers are derived later, but I want you to first learn how to perform the calculation. For now, the only year code you need to know is that 2000 has year code 0. Let's use this information to determine the day of the week of March 19 (my birthday) that year.

Since March has a month code of 2, and 2000 has a year code of 0, then our formula tells us that March 19, 2000, has

$$\text{Day of Week} = 2 + 19 + 0 = 21 \equiv 0 \ (\text{mod } 7)$$

Therefore, March 19, 2000, was a Sunday.

> ✖ **Aside**
>
> Here is a quick explanation of where the month codes come from. Notice that in a non-leap year, the month codes for February and March are the same. That makes sense, because when February has 28 days, then March 1 is 28 days after February 1, and so both months will begin on the same day of the week. Now as it happens, March 1, 2000, was a Wednesday. So if we want to give 2000 a year code of 0 and we want to give Monday a day code of 1, then that forces the month code for March to be 2. Thus, if it's not a leap year, then February must have a code of 2 too! And since March has 31 days, which is 3 greater than 28, then the April calendar is shifted 3 days further, which is why it has a month code of $2 + 3 = 5$. And when we add the $28 + 2$ days of April to the month code of 5, we see that May must have month code $5 + 2 = 7$, which can be reduced to 0 since we are working mod 7. Continuing this process, we can determine the month codes for the rest of the year.
>
> On the other hand, in a leap year (like 2000), February has 29 days, so the March calendar will be one day ahead of February's calendar, which is why the month code for February is $2 - 1 = 1$ in a leap year. January has 31 days, so its month code must be 3 below the month code for February. So in a non-leap year, the month code for January will be $2 - 3 = -1 \equiv 6 \ (\text{mod } 7)$. In a leap year, the month code for January will be $1 - 3 = -2 \equiv 5 \ (\text{mod } 7)$.

What happens to your birthday as you go from one year to the next? Normally, there are 365 days between your birthdays, and when that happens, your birthday advances by one day because $365 \equiv 1 \ (\text{mod } 7)$, since $365 = 52 \times 7 + 1$. But when February 29 appears between your birthdays (assuming you weren't born on February 29 yourself), then your birthday will advance by two days instead. In terms of our formula, we simply add 1 to the year code each year, except in leap years, when we add 2 instead. Here are the year codes for years 2000 through 2031. Don't worry. You will *not* need to memorize this!

Notice how the year codes begin $0, 1, 2, 3$, then at 2004 we leap over the 4 for a year code of 5. Then 2005 has year code 6, and 2006 should have year code 7, but since we are working mod 7, we simplify this number to 0. Then 2007 has year code 1, then 2008 (a leap year) has

Year	Code	Year	Code	Year	Code	Year	Code
2000*	0	2008*	3	2016*	6	2024*	2
2001	1	2009	4	2017	0	2025	3
2002	2	2010	5	2018	1	2026	4
2003	3	2011	6	2019	2	2027	5
2004*	5	2012*	1	2020*	4	2028*	0
2005	6	2013	2	2021	5	2029	1
2006	0	2014	3	2022	6	2030	2
2007	1	2015	4	2023	0	2031	3

Year codes from 2000 through 2031 (* denotes leap year)

year code 3, and so on. Using the table, we can determine that in 2025 (the next year which will be a perfect square), Pi Day (March 14) will be on

$$\text{Day of Week} = 2 + 14 + 3 = 19 \equiv 5 \ (\text{mod } 7) = \text{Friday}$$

How about January 1, 2008? Note that 2008 is a leap year, so the month code for January will be 5 instead of 6. Consequently, we have

$$\text{Day of Week} = 5 + 1 + 3 = 9 \equiv 2 \ (\text{mod } 7) = \text{Tuesday}$$

Notice that when you read across each row in the table, as we gain 8 years, our year code always increases by 3 (mod 7). For instance, the first row has $0, 3, 6, 2$ (where 2 is the same as 9 (mod 7)). That's because in any 8-year period, the calendar will always experience two leap years, so the dates will shift by $8 + 2 = 10 \equiv 3 \ (\text{mod } 7)$.

Here's even better news. Between 1901 and 2099, the calendar will repeat every 28 years. Why? In 28 years, we are guaranteed to experience exactly 7 leap years, so the calendar will shift by $28 + 7 = 35$ days, which leaves the day of the week unchanged, since 35 is a multiple of 7. (This statement is not true if the 28-year period crosses 1900 or 2100, since those years are not leap years.) Thus by adding or subtracting multiples of 28, you can turn *any* year between 1901 and 2099 into a year between 2000 and 2027. For example, 1983 has the same year code as $1983 + 28 = 2011$. The year 2061 has the same year code as $2061 - 56 = 2005$.

Thus, for all practical purposes, you can convert any year into one of the years on this table, and these year codes can be calculated pretty easily. For example, why should 2017 have a year code of 0? Well, starting in 2000, which has a year code of 0, the calendar has shifted 17 times *plus* an additional 4 times to account for the leap years of 2004, 2008, 2012, and 2016. Hence the year code for 2017 is $17 + 4 = 21 \equiv 0 \pmod 7$. How about 2020? This time we have 5 leap-year shifts (including 2020) so the calendar shifts $20 + 5 = 25$ times, and since $25 \equiv 4 \pmod 7$, the year 2020 has a year code of 4. In general, for any year between 2000 and 2027, you can determine its year code as follows.

Step 1: Take the last two digits of the year. For example, with 2022, the last two digits are 22.

Step 2: Divide that number by 4, and ignore any remainder. (Here, $22 \div 4 = 5$ with a remainder of 2.)

Step 3: Add the numbers in Steps 1 and 2. Here, $22 + 5 = 27$.

Step 4: Subtract the biggest multiple of 7 below the number in Step 3 (which will either be 0, 7, 14, 21, or 28) to obtain the year code. (In other words, reduce the number in Step 3, mod 7.) Since $27 - 21 = 6$, then the year code for 2022 is 6.

Note that Steps 1 through 4 will work for any year between 2000 and 2099, but the mental math is usually simpler if you first subtract a multiple of 28 to bring the year between 2000 and 2027. For example, the year 2040 can be first reduced to 2012, then Steps 1 through 4 produce a year code of $12 + 3 - 14 = 1$. But you can also work directly on 2040 to obtain the same year code: $40 + 10 - 49 = 1$.

The same steps can be applied to years outside of the 2000s. The month codes do not change. There is just one little adjustment for the year codes. The year code for 1900 is 1. Consequently, the codes from 1900 through 1999 are exactly one larger than their respective codes from 2000 through 2099. So since 2040 has a year code of 1, then 1940 will have a year code of 2. Since 2022 has a year code of 6, then 1922 will have a year code of 7 (or, equivalently, 0). The year 1800 has a year code of 3, 1700 has a year code of 5, and 1600 has a year code of 0. (In fact, the calendar will cycle every 400 years, since in 400 years it will have exactly $100 - 3 = 97$ leap years and so 400 years from now, the calendar will shift $400 + 97 = 497$ days, which is the same as today, since 497 is a multiple of 7.)

What day of the week was July 4, 1776? To find the year code for 2076, we first subtract 56 and then compute the year code for 2020: $20 + 5 - 21 = 4$. Thus the year code for 1776 is $4 + 5 = 9 \equiv 2 \pmod 7$.

Therefore, on the Gregorian calendar, July 4, 1776, has

$$\text{Day of Week} = 5 + 4 + 2 = 11 \equiv 4 \pmod{7} = \text{Thursday}$$

Perhaps the signers of the Declaration of Independence needed to pass legislation quickly before the long holiday weekend?

✂ Aside

Let's end this chapter with another magical property of the number 9. Take any number that has different digits, written from smallest to largest. Such numbers include 12345, 2358, 369, or 135789. Multiply this number by 9 and add the digits. Although we expect the sum to be a multiple of 9, it is quite surprising that the sum of the digits will always be *exactly* 9. For instance,

$$9 \times 12345 = 111,105 \quad 9 \times 2358 = 21,222 \quad 9 \times 369 = 3321$$

It even works if digits are repeated, as long as the number is written from smallest to largest and the ones digit is not equal to the tens digit. For example,

$$9 \times 12223 = 110,007 \quad 9 \times 33344449 = 300,100,041$$

So why does this work? Let's see what happens when we multiply 9 by the number $ABCDE$, where $A \leq B \leq C \leq D < E$. Since multiplying by 9 is the same as multiplying by $10 - 1$, this is the same as the subtraction problem

$$\begin{array}{r} A\ B\ C\ D\ E\ 0 \\ -\quad A\ B\ C\ D\ E \\ \hline \end{array}$$

If we do the subtraction from left to right, then since $B \geq A$ and $C \geq B$ and $D \geq C$ and $E > D$, this becomes the subtraction problem

$$\begin{array}{r} A\ (B\text{-}A)\ (C\text{-}B)\ (D\text{-}C)\ (E\text{-}D)\quad 0 \\ -\qquad\qquad\qquad\qquad\qquad\qquad E \\ \hline A\ (B\text{-}A)\ (C\text{-}B)\ (D\text{-}C)\ (E\text{-}D\text{-}1)\ (10\text{-}E) \end{array}$$

and the sum of the digits of our answer is

$$A + (B - A) + (C - B) + (D - C) + (E - D - 1) + (10 - E) = 9$$

as desired.

CHAPTER FOUR

$3! - 2! = 4$

The Magic of Counting

Math with an Exclamation Point!

We began this book with the problem of adding the numbers from 1 to 100. We discovered the total of 5050 and found a nice formula for the sum of the first n numbers. Now suppose we were interested in finding the *product* of the numbers from 1 to 100; what would we get? A really big number! If you're curious, it's the 158-digit number given below:

93326215443944152681699238856266700490715968264381621468592963895217599993229915608941463976156518286253697920827223758251185210916864000000000000000000000000000000000000

In this chapter, we will see how numbers like this are the foundation of counting problems. These numbers will enable us to determine such things as the number of ways to arrange a dozen books on a bookshelf (nearly half a *billion*), your chance of being dealt at least one pair in poker (not bad), and your chance of winning the lottery (not good).

When we multiply the numbers from 1 to n together, we denote the product as $n!$, which is pronounced "n factorial." In other words,

$$n! = n \times (n-1) \times (n-2) \times \cdots \times 3 \times 2 \times 1$$

For example,

$$5! = 5 \times 4 \times 3 \times 2 \times 1 = 120$$

71

I think the exclamation point is the appropriate notation since the number $n!$ grows very quickly and, as we'll see, it has many exciting and surprising applications. For convenience, mathematicians define $0! = 1$, and $n!$ is not defined when n is a negative number.

> **✕ Aside**
>
> From its definition, many people expect that $0!$ should be equal to 0. But let me try to convince you why $0! = 1$ makes sense. Notice that for $n \geq 2$, $n! = n \times (n-1)!$, and so
>
> $$(n-1)! = \frac{n!}{n}$$
>
> If we want that statement to remain true when $n = 1$, this would require that
>
> $$0! = \frac{1!}{1} = 1$$

As seen below, factorials grow surprisingly quickly:

$$
\begin{aligned}
0! &= 1 \\
1! &= 1 \\
2! &= 2 \\
3! &= 6 \\
4! &= 24 \\
5! &= 120 \\
6! &= 720 \\
7! &= 5040 \\
8! &= 40{,}320 \\
9! &= 362{,}880 \\
10! &= 3{,}628{,}800 \\
11! &= 39{,}916{,}800 \\
12! &= 479{,}001{,}600 \\
13! &= 6{,}227{,}020{,}800 \\
20! &= 2.43 \times 10^{18} \\
52! &= 8.07 \times 10^{67} \\
100! &= 9.33 \times 10^{157}
\end{aligned}
$$

How big are these numbers? It's been estimated that there are about 10^{22} grains of sand in the world and about 10^{80} atoms in the universe.

If you thoroughly mix up a deck of 52 cards, which we'll see can be done 52! ways, there is a very good chance that the order you obtained has never been seen before and will never be seen again, even if every person on earth produced a new shuffled deck every minute for the next million years!

✄ Aside

At the beginning of this chapter, you probably noticed that 100! ended with lots of zeros. Where did all the zeros come from? When multiplying the numbers from 1 to 100 we obtain a 0 every time a multiple of 5 is multiplied by a multiple of 2. Among the numbers from 1 to 100 there are 20 multiples of 5 and 50 even numbers, which might suggest that we should have 20 zeros at the end. But the numbers 25, 50, 75, and 100 each contribute an additional factor of 5, so 100! will end with 24 zeros.

Just like in Chapter 1, there are many beautiful number patterns using factorials. Here is one my favorites.

$$1 \cdot 1! = 1 = 2! - 1$$
$$1 \cdot 1! + 2 \cdot 2! = 5 = 3! - 1$$
$$1 \cdot 1! + 2 \cdot 2! + 3 \cdot 3! = 23 = 4! - 1$$
$$1 \cdot 1! + 2 \cdot 2! + 3 \cdot 3! + 4 \cdot 4! = 119 = 5! - 1$$
$$1 \cdot 1! + 2 \cdot 2! + 3 \cdot 3! + 4 \cdot 4! + 5 \cdot 5! = 719 = 6! - 1$$
$$\vdots$$

A factorial number pattern

The Rule of Sum and Product

Most counting problems essentially boil down to two rules: The rule of sum and the rule of product. The **rule of sum** is used when you want to count the total number of options you have when you have different types of choices. For example, if you had 3 short-sleeved shirts and 5 long-sleeved shirts, then you have 8 different choices for which shirt to wear. In general, if you have two types of objects, where there are a choices for the first type and b choices for the second type, then there are $a + b$ different objects (assuming that none of the b choices are the same as any of the a choices).

> ✗ **Aside**
>
> The rule of sum, as stated, assumes that the two types of objects have no objects in common. But if there are c objects that belong to both types, then those objects would be counted twice. Hence the number of different objects would be $a + b - c$. For example, if a class of students has 12 dog owners, 19 cat owners, and 7 students who own both dogs and cats, then the number of students who own a cat or a dog would be $12 + 19 - 7 = 24$. For a more mathematical example, among the numbers from 1 to 100 there are 50 multiples of 2, 33 multiples of 3, and 16 numbers that are multiples of 2 and 3 (namely the multiples of 6). Hence the number of numbers between 1 and 100 that are multiples of 2 or 3 is $50 + 33 - 16 = 67$.

The rule of product says that if an action consists of two parts, and there are a ways to do the first part and then b ways to do the second part, then the action can be completed in $a \times b$ ways. For instance, if I own 5 different pairs of pants and 8 different shirts, and if I don't care about color coordination (which I'm afraid applies to most mathematicians), then the number of different outfits is $5 \times 8 = 40$. If I own 10 ties and an outfit consists of shirt, pants, and tie, then there would be $40 \times 10 = 400$ outfits.

In a typical deck of cards, each card is assigned one of 4 suits (spades, hearts, diamonds, or clubs) and one of 13 values (A, 2, 3, 4, 5, 6, 7, 8, 9, 10, J, Q, or K). Hence the number of cards in a deck is $4 \times 13 = 52$. If we wanted to, we could deal out all 52 cards in a 4-by-13 rectangle, which is another way of seeing that there are 52 cards.

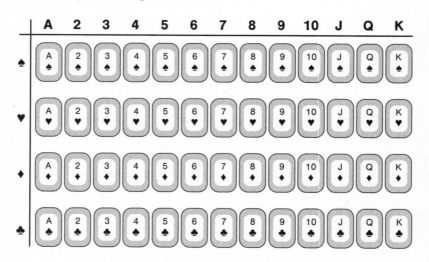

Let's apply the rule of product to count zip codes. How many five-digit zip codes are theoretically possible? Each digit of a zip code can be any number from 0 to 9, so the smallest zip code could be 00000 and the largest would be 99999; hence there are 100,000 possibilities. But you can also see this by the rule of product. You have 10 choices for the first digit (0 through 9) then 10 choices for the second digit, 10 choices for the third, 10 choices for the fourth, and 10 choices for the fifth. Hence there are $10^5 = 100,000$ possible zip codes.

When counting zip codes, numbers were allowed to be repeated. Now let's look at the situation where objects can't be repeated, such as when you are arranging objects in a row. It's easy to see that two objects can be arranged 2 ways. For instance, letters A and B can be arranged either as AB or as BA. And three objects can be arranged 6 ways: ABC, ACB, BAC, BCA, CAB, CBA. Now can you see how four objects can be arranged in 24 ways without explicitly writing them down? There are 4 choices for which letter goes first (either A or B or C or D). Once that letter has been chosen, there will be 3 choices for the next letter, then 2 choices for the next letter, and finally there is only 1 possibility for the last letter. Altogether, there are $4 \times 3 \times 2 \times 1 = 4! = 24$ possibilities. In general, *there are n! ways to arrange n different objects.*

We combine the rules of sum and product in this next example. Suppose a state manufactures license plates of two varieties. Type I license plates consist of 3 letters followed by 3 digits. Type II license plates consist of 2 letters followed by 4 digits. How many license plates are possible? (We allow all 26 letters and all 10 digits, ignoring the confusion that can occur with similar characters like O and 0.) From the rule of product, the number of Type I plates is:

$$26 \times 26 \times 26 \times 10 \times 10 \times 10 = 17,576,000$$

The number of Type II plates is

$$26 \times 26 \times 10 \times 10 \times 10 \times 10 = 6,760,000$$

Since every plate is either Type I *or* Type II (not both), then the rule of sum says that the number of possibilities is their total: 24,336,000.

One of the pleasures of doing counting problems (mathematicians call this branch of mathematics *combinatorics*) is that quite often you can solve the same problem in more than one way. (We saw this was true with mental arithmetic problems too.) The last problem can actually be done in one step. The number of license plates is

$$26 \times 26 \times 36 \times 10 \times 10 \times 10 = 24,336,000$$

since the first two characters of the plate can each be chosen 26 ways, the last 3 characters can each be chosen 10 ways, and the third character, depending on whether it was a letter or a digit, can be chosen $26 + 10 = 36$ ways.

Lotteries and Poker Hands

In this section, we'll apply our new counting skills to determine the chance of winning the lottery or being dealt various types of poker hands. But first let's relax with a little bit of ice cream.

Suppose an ice cream shop offers 10 flavors of ice cream. How many ways can you create a triple cone? When creating a cone, the order of the flavors matters (of course!). If the flavors are allowed to be repeated, then since we have 10 choices for each of the three scoops, there would be $10^3 = 1000$ possible cones. If we insist that all three flavors be different, then the number of cones is $10 \times 9 \times 8 = 720$, as illustrated opposite.

But now for the real question. How many ways can you put three *different* flavors in a cup, where order does *not* matter? Since order doesn't matter, there are fewer possibilities. In fact, there are 1/6th as many ways. Why is that? For any choice of three different flavors in a cup (say chocolate, vanilla, and mint chip), there are $3! = 6$ ways that they can be arranged on a cone. Hence there would be 6 times as many cones as cups. Consequently, the number of cups is

$$\frac{10 \times 9 \times 8}{3 \times 2 \times 1} = \frac{720}{6} = 120$$

Another way to write $10 \times 9 \times 8$ is $10!/7!$ (although the first expression is easier to calculate). Hence the number of cups can be expressed as $\frac{10!}{3!7!}$. We call this expression "10 choose 3," and it is denoted by the symbol $\binom{10}{3}$, which equals 120. In general, the number of ways to choose k different objects where order does not matter from a collection of n different objects is pronounced "n choose k" and has the formula

$$\binom{n}{k} = \frac{n!}{k!(n-k)!}$$

Mathematicians refer to these counting problems as *combinations*, and numbers of the form $\binom{n}{k}$ are called *binomial coefficients*. Counting problems where the order does matter are called *permutations*. It's easy to mix these terms up—for example, we often refer to a "combination

For every cup with 3 flavors, there are 3! = 6 ways to arrange them on a cone

lock" when we really should be saying "permutation lock," since the order that the numbers are used is important.

If the ice cream shop offered 20 flavors and you wish to fill a bucket with 5 scoops consisting of all different flavors (where order is not important), then there would be

$$\binom{20}{5} = \frac{20!}{5!15!} = \frac{20 \times 19 \times 18 \times 17 \times 16}{5!} = 15{,}504$$

possibilities. By the way, if your calculator does not have a button that computes $\binom{20}{5}$, you can just type the words "20 choose 5" into your favorite search engine and it will probably display a calculator with the answer.

Binomial coefficients sometimes appear in problems where order seems to matter. If we flip a coin 10 times, how many possible sequences are there (like HTHTTHHTTT or HHHHHHHHHH)? Since there are 2 possible outcomes for each flip, then the rule of product tells us that there are $2^{10} = 1024$ sequences, each of which has the same probability of occurring. (Some people are initially surprised by this, since the second sequence in our example may seem less probable than the first one, but they each have probability of $\frac{1}{1024}$.) On the other hand, it is way more likely for ten flips of a coin to produce 4 heads than 10 heads. There is just one way to achieve 10 heads, so that has probability $\frac{1}{1024}$. But how many ways can we get 4 heads out of 10? Such a sequence is determined by picking 4 of the 10 flips to be heads, and the rest are forced to be tails. The number of ways to determine *which* 4 of the 10 flips are to be heads is $\binom{10}{4} = 210$. (It's like picking 4 different scoops of ice cream among 10 possible flavors.) Hence when a fair coin is flipped 10 times, the probability that we get exactly 4 heads is

$$\frac{\binom{10}{4}}{2^{10}} = \frac{210}{1024}$$

which is about 20 percent of the time.

✂ Aside

It's natural to ask, from 10 possible flavors, how many cups with 3 scoops can be made if repetition of flavors is allowed? (The answer is not $10^3/6$, which is not even a whole number!) The direct approach would be to consider three cases, depending on the number of different flavors in the cup. Naturally there are 10 cups that use just one flavor and, from the above discussion, there are $\binom{10}{3} = 120$ cups that use three flavors. There are $2 \times \binom{10}{2} = 90$ cups that use two flavors, since we can choose the two flavors in $\binom{10}{2}$ ways, then decide which of the flavors gets two scoops. Adding all three cases together, the number of cups is $10 + 120 + 90 = 220$.

There is another way to get this answer without breaking it into three cases. Any cup can be represented using 3 stars and 9 bars. For example, choosing flavors 1, 2, and 2 can be represented by the star-bar arrangement

$$* \mid ** \mid \mid \mid \mid \mid \mid \mid \mid \mid$$

Picking flavors 2, 2, and 7 would look like

$$\mid ** \mid \mid \mid \mid \mid * \mid \mid \mid$$

and the star-bar arrangement

$$\mid \mid * \mid \mid * \mid \mid \mid \mid \mid *$$

would be a cup with flavors 3, 5, and 10. Every arrangement of 3 stars and 9 bars corresponds to a different cup. Together the stars and bars occupy 12 spaces, 3 of which are occupied by stars. Hence, the stars and bars can be arranged in $\binom{12}{3} = 220$ ways. More generally, the number of ways to choose k objects out of n where order is not important, but repetition is allowed, is the number of ways to arrange k stars and $n - 1$ bars, which can be done in $\binom{n+k-1}{k}$ ways.

05 08 13 21 34
MEGA 03

Many problems involving games of chance involve combinations. For example, in the California Lottery, you pick 5 different numbers between 1 and 47. Additionally, you choose a MEGA number between 1 and 27 (which is allowed to be a repeat of one of your 5 chosen numbers). There are 27 choices for the MEGA number and the other 5 numbers can be chosen $\binom{47}{5}$ ways. Hence the number of possibilities is

$$27 \times \binom{47}{5} = 41{,}416{,}353$$

Thus your chance of winning the lottery grand prize is about 1 in 40 million.

Now let's switch gears and consider the game of poker. A typical poker hand consists of 5 different cards chosen from 52 different cards, where the order of the 5 cards is unimportant. Hence the number of poker hands is

$$\binom{52}{5} = \frac{52!}{5!47!} = 2{,}598{,}960$$

In poker, five cards of the same suit, such as

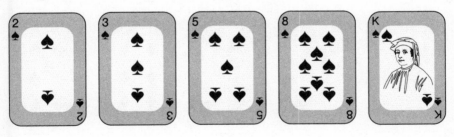

is called a *flush*. How many flushes are there? To create a flush, first choose your suit, which can be done in 4 ways. (Mentally, I like to commit to a definite choice, say spades.) Now how many ways can you choose 5 cards of that suit? From the 13 spades in a deck, 5 of them can be chosen in $\binom{13}{5}$ ways. Hence the number of flushes is

$$4 \times \binom{13}{5} = 5148$$

Hence the chance of being dealt a flush in poker is 5148/2,598,960, which is approximately 1 in 500. For poker purists, you can subtract $4 \times 10 = 40$ hands from the 5148 for *straight flushes*, when the flush uses five consecutive cards.

A *straight* in poker consists of 5 cards of consecutive values, such as A2345 or 23456 or ... or 10JQKA, like the one below.

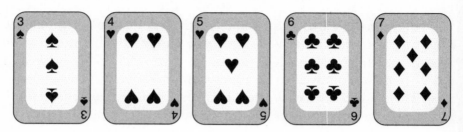

There are 10 different types of straights (defined by the lowest card), and once we choose its type (say 34567) then each of the 5 cards can be assigned one of 4 suits. Hence the number of straights is

$$10 \times 4^5 = 10{,}240$$

which is nearly twice as many as flushes. So the probability of being dealt a straight is about 1 in 250. That's the reason flushes are worth more than straights in poker: they are harder to get.

An even more valuable hand is the *full house*, consisting of 3 cards of one value and 2 cards of a different value. A typical hand might look like this:

To construct a full house, we first need to choose a value to be tripled (13 ways), then a value to be doubled (12 ways). (Say we've decided on using three queens and two 7s.) Then we need to assign suits. We can decide which three queens to use in $\binom{4}{3} = 4$ ways and which two 7s in $\binom{4}{2} = 6$ ways. Altogether the number of full houses is

$$13 \times 12 \times 4 \times 6 = 3744$$

So the probability of being dealt a full house is 3744/2,598,960, which is about 1 in 700.

Let's contrast the full house with getting exactly *two pair*. Such hands have two cards of one value, two cards of a different value, and another card of a third value, like

Many people mistakenly start counting the paired values with 13×12, but this double-counts, since first picking queens followed by 7s is the same as first picking 7s followed by queens. The correct way to start is $\binom{13}{2}$ (say picking queens and 7s together), then choosing a new value for the unpaired card (like 5), then assign suits. The number of two-pair hands is

$$\binom{13}{2}\binom{11}{1}\binom{4}{2}\binom{4}{2}\binom{4}{1} = 123{,}552$$

which occurs about 5 percent of the time.

We won't go through all of the rest of the poker hands in detail, but see if you can verify the following. The number of poker hands that are *four of a kind*, like $A\spadesuit A\heartsuit A\diamondsuit A\clubsuit 8\diamondsuit$, is

$$\binom{13}{1}\binom{12}{1}\binom{4}{4}\binom{4}{1} = 13 \times 12 \times 1 \times 4 = 624$$

A poker hand like $A\spadesuit A\heartsuit A\diamondsuit 9\clubsuit 8\diamondsuit$ is called *three of a kind*. The number of these is

$$\binom{13}{1}\binom{12}{2}\binom{4}{3}\binom{4}{1}\binom{4}{1} = 54{,}912$$

The number of poker hands with exactly *one pair*, like $A\spadesuit A\heartsuit J\diamondsuit 9\clubsuit 8\diamondsuit$, is

$$\binom{13}{1}\binom{12}{3}\binom{4}{2}4^3 = 1{,}098{,}240$$

about 42 percent of all hands.

> ✗ **Aside**
>
> So how many hands might be classified as *junk*, containing no pairs, no straight, and no flush? You could carefully add up the cases above and subtract from $\binom{52}{5}$, but here is a direct answer:
>
> $$\left(\binom{13}{5} - 10 \right) (4^5 - 4) = 1{,}302{,}540$$
>
> The first term counts the ways to choose any 5 different values (preventing two or more cards of the same value) with the exception of the 10 ways of picking five consecutive values like 34567. Then the next term assigns a suit to each of these five different values; there are 4 choices for each value, but we have to throw away the 4 possibilities that they were all assigned the same suit. The upshot is that about 50.1 percent of all hands are worth less than one pair. But that means 49.9 percent are worth one pair or better.

Here's a question that has three interesting answers, two of which are actually correct! How many five-card hands contain at least one ace? A tempting *wrong* answer is simply $4 \times \binom{51}{4}$. The (faulty) reasoning is that you pick an ace 4 ways, and then the other 4 cards can be chosen freely among the remaining 51 cards (including other aces). The problem with this reasoning is that some hands (those with more than one ace) get counted more than once. For instance, the hand $A\spadesuit A\heartsuit J\diamondsuit 9\clubsuit 8\diamondsuit$ would get counted when we first choose $A\spadesuit$ (then the other 4 cards) as well as when we first choose the $A\heartsuit$ (then the other 4 cards). A correct way to handle this problem is to break the problem into four cases, depending on the number of aces in the hand. For instance, the number of hands with exactly one ace is $\binom{4}{1}\binom{48}{4}$ (by picking one ace, then four non-aces). If we continue this way, and count hands with two, three, or four aces, the total number of hands with at least one ace is

$$\binom{4}{1}\binom{48}{4} + \binom{4}{2}\binom{48}{3} + \binom{4}{3}\binom{48}{2} + \binom{4}{4}\binom{48}{1} = 886{,}656$$

But a quicker calculation can be done by addressing the *opposite* question. The number of hands with *no aces* is simply $\binom{48}{5}$. Hence the number of hands with at least one ace is

$$\binom{52}{5} - \binom{48}{5} = 886{,}656$$

We noted earlier that poker hands are ranked according to how rare they are. For example, since there are more ways to be a single pair than two pair, one pair is worth less than two pair. The order of poker hands, from lowest to highest, is:

One pair
Two pair
Three of a kind
Straight
Flush
Full house
Four of a kind
Straight flush

A simple way to remember this is "One, two, three, straight, flush; two-three, four, straight-flush." ("Two-three" is the full house.)

Now suppose we play poker with jokers (cards, not people). Here we have 54 cards where the two jokers are wild and can be assigned any value that gives you the best hand. So, for example, if you end up with A$\heartsuit$, A$\diamondsuit$, K$\spadesuit$, 8$\diamondsuit$ and joker, you would choose to make the joker an ace to give yourself three aces. If you turn the joker into a king, then your hand would be two pair, which is an inferior hand.

What card should we assign the joker to achieve the best hand?

But here's where things get interesting. Under the traditional ordering of hands, if you are presented with hands like the one above that could be assigned two pair or three of a kind, then you would count it as three of a kind instead of two pair. But the consequence of this is that there will be more hands that would count as three of a kind than as two pair, so two pair would become the rarer hand. But if we try to fix that problem by making the two-pair hand more valuable, then the same problem occurs and we wind up with more hands that would count as two pair than as three of a kind. The surprising conclusion, discovered in 1996 by mathematician Steve Gadbois, is that when you play poker with jokers, there is no consistent way to rank the hands in order of frequency.

Patterns in Pascal's Triangle

Behold Pascal's triangle:

Row 0:	$\binom{0}{0}$
Row 1:	$\binom{1}{0}$ $\binom{1}{1}$
Row 2:	$\binom{2}{0}$ $\binom{2}{1}$ $\binom{2}{2}$
Row 3:	$\binom{3}{0}$ $\binom{3}{1}$ $\binom{3}{2}$ $\binom{3}{3}$
Row 4:	$\binom{4}{0}$ $\binom{4}{1}$ $\binom{4}{2}$ $\binom{4}{3}$ $\binom{4}{4}$
Row 5:	$\binom{5}{0}$ $\binom{5}{1}$ $\binom{5}{2}$ $\binom{5}{3}$ $\binom{5}{4}$ $\binom{5}{5}$
Row 6:	$\binom{6}{0}$ $\binom{6}{1}$ $\binom{6}{2}$ $\binom{6}{3}$ $\binom{6}{4}$ $\binom{6}{5}$ $\binom{6}{6}$
Row 7:	$\binom{7}{0}$ $\binom{7}{1}$ $\binom{7}{2}$ $\binom{7}{3}$ $\binom{7}{4}$ $\binom{7}{5}$ $\binom{7}{6}$ $\binom{7}{7}$
Row 8:	$\binom{8}{0}$ $\binom{8}{1}$ $\binom{8}{2}$ $\binom{8}{3}$ $\binom{8}{4}$ $\binom{8}{5}$ $\binom{8}{6}$ $\binom{8}{7}$ $\binom{8}{8}$
Row 9:	$\binom{9}{0}$ $\binom{9}{1}$ $\binom{9}{2}$ $\binom{9}{3}$ $\binom{9}{4}$ $\binom{9}{5}$ $\binom{9}{6}$ $\binom{9}{7}$ $\binom{9}{8}$ $\binom{9}{9}$
Row 10:	$\binom{10}{0}$ $\binom{10}{1}$ $\binom{10}{2}$ $\binom{10}{3}$ $\binom{10}{4}$ $\binom{10}{5}$ $\binom{10}{6}$ $\binom{10}{7}$ $\binom{10}{8}$ $\binom{10}{9}$ $\binom{10}{10}$

Pascal's triangle with symbols

In Chapter 1, we saw interesting patterns arise when we put numbers in triangles. The numbers $\binom{n}{k}$ that we've just been studying contain their own beautiful patterns when viewed in a triangle, called *Pascal's triangle*, as displayed above. Using our formula $\binom{n}{k} = \frac{n!}{k!(n-k)!}$, let's turn these symbols into numbers and look for patterns (see next page). We will explain most of these patterns in this chapter, but feel free to skip the explanations on first reading and just enjoy the patterns.

The top row (called row 0) has just one term, namely $\binom{0}{0} = 1$. (Remember $0! = 1$.) Every row will begin and end with 1 since

$$\binom{n}{0} = \frac{n!}{0!n!} = 1 = \binom{n}{n}$$

Have a look at row 5.

Row 5: 1 5 10 10 5 1

Notice that the second entry is 5, and in general, the second entry of row n is n. This makes sense because $\binom{n}{1}$, the number of ways to choose one object from n objects, is equal to n. Notice also that each row is

Row 0:										1									
Row 1:									1		1								
Row 2:								1		2		1							
Row 3:							1		3		3		1						
Row 4:						1		4		6		4		1					
Row 5:					1		5		10		10		5		1				
Row 6:				1		6		15		20		15		6		1			
Row 7:			1		7		21		35		35		21		7		1		
Row 8:		1		8		28		56		70		56		28		8		1	
Row 9:	1		9		36		84		126		126		84		36		9		1
Row 10:	1	10		45		120		210		252		210		120		45		10	1

Pascal's triangle with numbers

symmetric: it reads the same backward as it does forward. For example, in row 5, we have

$$\binom{5}{0} = 1 = \binom{5}{5}$$

$$\binom{5}{1} = 5 = \binom{5}{4}$$

$$\binom{5}{2} = 10 = \binom{5}{3}$$

In general, the pattern says that

$$\binom{n}{k} = \binom{n}{n-k}$$

✗ Aside

This symmetry relationship can be justified two ways. From the formula, we can algebraically show that

$$\binom{n}{n-k} = \frac{n!}{(n-k)!(n-(n-k))!} = \frac{n!}{(n-k)!k!} = \binom{n}{k}$$

But you don't really need the formula to see why it is true. For example, why should $\binom{10}{3} = \binom{10}{7}$? The number $\binom{10}{3}$ counts the ways to choose 3 ice cream flavors (from 10 possibilities) to put in a cup. But this is the same as choosing 7 flavors to *not* put in the cup.

The next pattern that you might notice is that, except for the 1s at the beginning and end of the row, every number is the sum of the two numbers above it. This relationship is so striking that we call it *Pascal's identity*. For example, look at rows 9 and 10 in Pascal's triangle.

Row 9:   1   9   ⑨36   ⑧84   126 126   84   36   9   1

Row 10:   1   10   45   ⑫120   210 252 210 120   45   10   1

Each number is the sum of the two numbers above it

Now why should that be? When we see that $120 = 36 + 84$, this is a statement about the counting numbers

$$\binom{10}{3} = \binom{9}{2} + \binom{9}{3}$$

To see why this is true, let's ask the question: If an ice cream store sells 10 flavors of ice cream, how many ways can you choose 3 different flavors of ice cream for a cup (where order is not important)? The first answer, as we have already noted, is $\binom{10}{3}$. But there is another way to answer the question. Assuming that one of the possible flavors is vanilla, how many of the cups do *not* contain vanilla? That would be $\binom{9}{3}$, since we can choose any 3 flavors from the remaining 9 flavors. How many cups *do* contain vanilla? If vanilla is required to be one of the flavors, then there are just $\binom{9}{2}$ ways to choose the remaining two flavors in the cup. Hence the number of possible cups is $\binom{9}{2} + \binom{9}{3}$. Which answer is right? Our logic was correct both times, so they are both right, and hence the two quantities are equal. By the same logic (or algebra, if you prefer), it follows that for any number k between 0 and n,

$$\binom{n}{k} = \binom{n-1}{k-1} + \binom{n-1}{k}$$

Now let's see what happens when we add all of the numbers in each row of Pascal's triangle, as seen on the opposite page.

The pattern suggests that the sum of the numbers in each row is always a power of 2. Specifically row n will add up to 2^n. Why should that be? Another way to describe the pattern is that the first row sums to 1, and then the sum doubles with each new row. This makes sense if you think of Pascal's identity, which we just proved. For instance, when we add the numbers in row 5, and rewrite them in terms of the numbers in row 4, we get

$$1 + 5 + 10 + 10 + 5 + 1$$

$$1 \qquad\qquad = 1$$

$$1 + 1 \qquad\qquad = 2$$

$$1 + 2 + 1 \qquad\qquad = 4$$

$$1 + 3 + 3 + 1 \qquad\qquad = 8$$

$$1 + 4 + 6 + 4 + 1 \qquad\qquad = 16$$

$$1 + 5 + 10 + 10 + 5 + 1 \qquad = 32$$

$$\vdots$$

In Pascal's triangle, the row sums are powers of 2

$$= 1 + (1 + 4) + (4 + 6) + (6 + 4) + (4 + 1) + 1$$
$$= (1 + 1) + (4 + 4) + (6 + 6) + (4 + 4) + (1 + 1)$$

which is literally twice the sum of the numbers in row 4. And by the same reasoning, this doubling pattern will continue forever.

In terms of the binomial coefficients, this identity says that when we sum the numbers in row n:

$$\binom{n}{0} + \binom{n}{1} + \binom{n}{2} + \cdots + \binom{n}{n} = 2^n$$

which is somewhat surprising, since the individual terms are evaluated with factorials, and are often divisible by many different numbers. And yet the grand total has 2 as its only prime factor.

Another way to explain this pattern is through counting. We call such an explanation a *combinatorial proof*. To explain the sum of the numbers in row 5, let's go to an ice cream store that offers 5 flavors. (The argument for row n is similar.) How many ways can we put different scoops of ice cream in our cup with the restriction that no flavor is allowed to be repeated? Our cup is allowed to have 0 or 1 or 2 or 3 or 4 or 5 different flavors, and the order of the scoops in the cup is not important. How many ways can it have exactly 2 scoops? As we've seen before, this can be done $\binom{5}{2} = 10$ different ways. Altogether, depending

How many ways can we put different scoops of ice cream in our cup?

on the number of scoops in the cup, the rule of sum tells us that there are

$$\binom{5}{0} + \binom{5}{1} + \binom{5}{2} + \binom{5}{3} + \binom{5}{4} + \binom{5}{5}$$

ways, which simplifies to $1 + 5 + 10 + 10 + 5 + 1$. On the other hand, we can also answer our question with the rule of product. Instead of deciding in advance how many scoops will be in our cup, we can look at each flavor and decide, yes or no, whether it will be in the cup. For instance, we have 2 choices for chocolate (yes or no), then 2 choices for vanilla (yes or no), and so on down to the fifth flavor. (Notice that if we decide "no" on each flavor, then we have an empty cup, which is permissible.) Hence the number of ways we can make our decision is

$$2 \times 2 \times 2 \times 2 \times 2 = 2^5$$

Since our reasoning was correct both times, it follows that

$$\binom{5}{0} + \binom{5}{1} + \binom{5}{2} + \binom{5}{3} + \binom{5}{4} + \binom{5}{5} = 2^5$$

as predicted.

✂ **Aside**

A similar combinatorial argument shows that if we sum *every other* number in row n, we get a total of 2^{n-1}. This is no surprise in odd-numbered rows like row 5, since the numbers we are adding, $1 + 10 + 5$, are the same as the numbers we are excluding, $5 + 10 + 1$, so we get half of the 2^n grand total. But it also works in even-numbered rows as well. For instance, in row 4, $1 + 6 + 1 = 4 + 4 = 2^3$. In general, for any row $n \geq 1$, we have

$$\binom{n}{0} + \binom{n}{2} + \binom{n}{4} + \binom{n}{6} + \cdots = 2^{n-1}$$

Why? The left side counts those ice cream cups that have an even number of scoops (when there are n possible flavors and all scoops must be different flavors). But we can also create such a cup by freely choosing flavors 1 through $n - 1$. We have 2 choices for the first flavor (yes or no), 2 choices for the second flavor, ..., and 2 choices for the $(n - 1)$st flavor. But there is just one choice for the last flavor if we want the total number of flavors to be even. Hence the number of even-sized cups is 2^{n-1}.

When we write Pascal's triangle as a *right triangle*, more patterns emerge. The first column (column 0) consists of all 1s, the second column (column 1) are the positive integers $1, 2, 3, 4$, and so on. Column 2, beginning $1, 3, 6, 10, 15 \ldots$, should also look familiar. These are the triangular numbers that we saw in Chapter 1. In general, the numbers in column 2 can also be expressed as

$$\binom{2}{2}, \binom{3}{2}, \binom{4}{2}, \binom{5}{2}, \binom{6}{2}, \ldots$$

and column k consists of the numbers $\binom{k}{k}, \binom{k+1}{k}, \binom{k+2}{k}$, and so on.

Now look at what happens when you add the first few (or many) numbers of any column. For instance, if we add the first 5 numbers of column 2 (see next page), we get $1 + 3 + 6 + 10 + 15 = 35$, which happens to be the number diagonally down from it. In other words:

$$\binom{2}{2} + \binom{3}{2} + \binom{4}{2} + \binom{5}{2} + \binom{6}{2} = \binom{7}{3}$$

This is an example of the *hockey stick identity* because the pattern formed in Pascal's triangle resembles a hockey stick, with a long column of numbers, next to another number jutting out from it. To understand why this pattern works, imagine we have a hockey team with 7 players, each with a different number on their jersey: 1, 2, 3, 4, 5, 6, 7. How

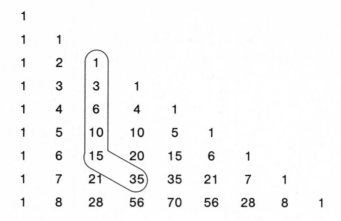

Pascal's right triangle displays a "hockey stick" pattern

many ways can I choose 3 of these players for a practice session? Since order is not important, there are $\binom{7}{3}$ ways this can be done. Now let's answer that same question by breaking it into cases. How many of these ways include player 7? Equivalently, this asks: How many have 7 as the largest jersey number? Since 7 is included, there are $\binom{6}{2}$ ways to pick the other two players. Next, how many of these ways have 6 as the largest jersey number? Here, 6 must be chosen, and 7 must not be chosen, so there are $\binom{5}{2}$ ways to pick the other two players. Likewise, there are $\binom{4}{2}$ ways for 5 to be the largest number, $\binom{3}{2}$ ways for 4 to be the largest number, and $\binom{2}{2} = 1$ way for 3 to be the largest number. Since the largest of the three numbers must be 3, 4, 5, 6, or 7, we have counted all of the possibilities, so the three members can be chosen in $\binom{2}{2} + \binom{3}{2} + \cdots \binom{6}{2}$ ways, as described by the left side of the previous equation. More generally, this argument shows that

$$\binom{k}{k} + \binom{k+1}{k} + \cdots + \binom{n}{k} = \binom{n+1}{k+1}$$

Let's apply this formula to resolve an important problem that you probably think about every year during the holiday season. According to the popular song "The Twelve Days of Christmas," on the first day your true love gives you 1 gift (a partridge). On the second day you get 3 gifts (1 partridge and 2 turtledoves). On the third day you get 6 gifts (1 partridge, 2 turtledoves, 3 French hens), and so on. The question is: After the 12th day, how many gifts do you receive in total?

After the 12th day of Christmas, how many gifts did my true love give to me?

On the nth day of Christmas, the number of gifts you receive is

$$1 + 2 + 3 + \cdots + n = \frac{n(n+1)}{2} = \binom{n+1}{2}$$

(This comes from our handy formula for triangular numbers or from the hockey stick identity when $k = 1$.) So on the first day you get $\binom{2}{2} = 1$ gift; on the second day you get $\binom{3}{2} = 3$ gifts, and so on up through the 12th day, when you receive $\binom{13}{2} = \frac{13 \times 12}{2} = 78$ gifts. Applying the hockey stick identity, it follows that the total number of gifts will be

$$\binom{2}{2} + \binom{3}{2} + \cdots + \binom{13}{2} = \binom{14}{3} = \frac{14 \times 13 \times 12}{3!} = 364$$

Thus, if you spread these gifts out over the next year, you could enjoy a new gift practically every day (perhaps taking one day off for your birthday)!

Let's celebrate our answer to the last problem with a festive song I call "The nth Day of Christmas."

On the nth day of Christmas, my true love gave to me
n novel knick-knacks
$n - 1$ thing or other
$n - 2$ et cetera
$\vdots$

5 (plus 10) other things!
Counting all the gifts
Through day n,
What's the total sum?
It's precisely $\binom{n+2}{3}$.

Now here is one of the *oddest* patterns in Pascal's triangle. If you look at the triangle, where we have circled each of the odd numbers, you will see triangles within the triangle.

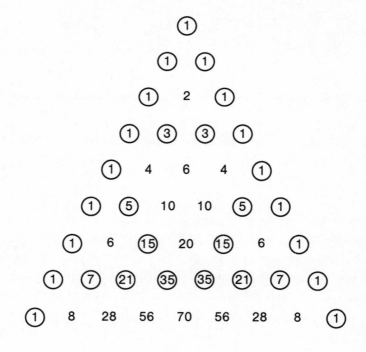

Odd numbers in Pascal's triangle

Now let's take a longer view of the triangle with 16 rows, where we replace each odd number with 1 and each even number with 0. Notice that underneath each pair of 0s and each pair of 1s, you get 0. This reflects the fact that you get an even total when you add two even numbers together or two odd numbers together.

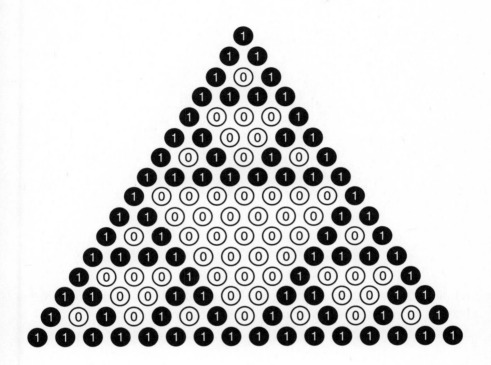

A longer view of the odd numbers

We have an even longer view of the triangle on the next page, consisting of the first 256 rows, where we have replaced each odd number with a black square and each even number with a white square.

This figure is an approximation of the fractal image known as the *Sierpinski triangle*. It's just one of the many hidden treasures inside Pascal's triangle. Here's another surprise. How many odd numbers are in each row of Pascal's triangle? Looking at rows 1 through 8 (excluding row 0), we count 2, 2, 4, 2, 4, 4, 8, 2, and so on. No obvious pattern jumps out, although it appears that the answer is always a power of 2. In fact, powers of 2 play an important role here. For example, notice that the rows with exactly 2 odd numbers are rows 1, 2, 4, and 8, which are powers of 2. For the general pattern, we exploit the fact that every

Pascal meets Sierpinski

whole number greater than or equal to zero can be obtained in a unique way by summing *distinct powers of* 2. For example:

$$
\begin{aligned}
1 &= 1 \\
2 &= 2 \\
3 &= 2+1 \\
4 &= 4 \\
5 &= 4+1 \\
6 &= 4+2 \\
7 &= 4+2+1 \\
8 &= 8
\end{aligned}
$$

We have 2 odd numbers in rows 1, 2, 4, and 8 (which are powers of 2). We have 4 odd numbers in rows 3, 5, and 6 (which are the sum of 2 powers of 2), and we have 8 odd numbers in row 7 (which is the sum of 3 powers of 2). Here is the surprising and beautiful rule. If n is the sum of p different powers of 2, then the number of odd numbers in row n is 2^p. So, for example, how many odd numbers are in row 83? Since

$83 = 64 + 16 + 2 + 1$, the sum of 4 powers of 2, then row 83 will have $2^4 = 16$ odd numbers!

> ## ✕ Aside
>
> We won't prove the fact here, but in case you're curious, $\binom{83}{k}$ is odd whenever
>
> $$k = 64a + 16b + 2c + d$$
>
> where a, b, c, d can be 0 or 1. Specifically, k must be equal to one of these numbers:
>
> $$0, 1, 2, 3, 16, 17, 18, 19, 64, 65, 66, 69, 80, 81, 82, 83$$

We end this chapter with one last pattern. We have seen what happens when we add along the rows of Pascal's triangle (powers of 2) and the columns of Pascal's triangle (hockey stick). But what happens when we add the diagonals?

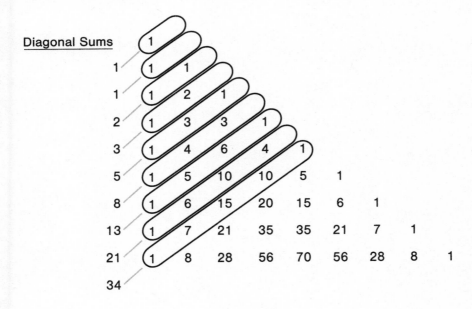

Pascal meets Fibonacci

When we add the diagonals, as illustrated above, we get the following totals:

$$1, 1, 2, 3, 5, 8, 13, 21, 34$$

These are the fabulous Fibonacci numbers, and they will be the subject of our next chapter.

CHAPTER FIVE

1, 1, 2, 3, **5**, 8, 13, 21...

The Magic of Fibonacci Numbers

Nature's Numbers

Behold one of the most magical sequences of numbers, the Fibonacci numbers!

1, 1, 2, 3, 5, 8, 13, 21, 34, 55, 89, 144, 233, . . .

The Fibonacci sequence starts with numbers 1 and 1. The third number is $1 + 1$ (the sum of the two previous numbers), which is 2. The fourth number is $1 + 2 = 3$, the fifth number is $2 + 3 = 5$, and the numbers continue to grow in leapfrog fashion: $3 + 5 = 8$, $5 + 8 = 13$, $8 + 13 = 21$, and so on. These numbers appeared in the book *Liber Abaci* in 1202 by Leonardo of Pisa (later nicknamed "Fibonacci"). *Liber Abaci*, which literally means "The Book of Calculation," introduced to the Western world the Indo-Arabic numerals and methods of arithmetic that we currently use today.

One of the book's many arithmetic problems involved immortal rabbits. Suppose that baby rabbits take one month to mature, and each pair produces a new pair of baby rabbits every month thereafter for the rest of their never-ending lives. The question is, if we start with one pair of baby rabbits, how many pairs of rabbits will there be twelve months later?

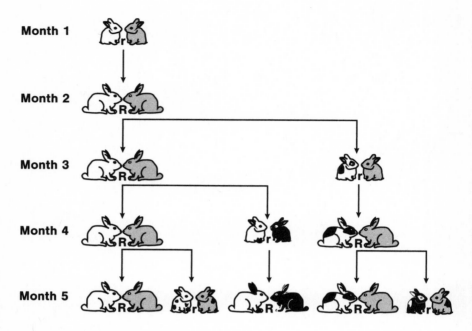

The problem can be illustrated through pictures or symbols. We let a lower-case letter "r" denote a baby pair of rabbits and an upper-case "R" denote an adult pair of rabbits. As we go from one month to the next, every little "r" becomes a big "R," and each "R" is replaced by "R r." (That is, little rabbits become big rabbits and big rabbits produce little rabbits.)

The situation can be modeled as in the following table. We see that in the first six months, the number of rabbit pairs is, respectively, 1, 1, 2, 3, 5, and 8.

Month Number	Population	Number of Rabbit Pairs
1	r	1
2	R	1
3	Rr	2
4	Rr R	3
5	Rr R Rr	5
6	Rr R Rr Rr R	8

Let's try to convince ourselves that there will be 13 pairs of rabbits in the seventh month, without explicitly listing the population. How many adult rabbit pairs will be alive in the seventh month? Since every rabbit that was alive in month six is now an adult rabbit, then there will be 8 adult rabbits in month seven.

How many baby rabbit pairs will be alive in the seventh month? That's equal to the number of adult rabbit pairs from the sixth month, namely 5, which is (not coincidentally) the same as the total population of the fifth month. Consequently, the number of rabbit pairs in the seventh month will be $8 + 5 = 13$.

If we call the first two Fibonacci numbers $F_1 = 1$ and $F_2 = 1$, then define the next Fibonacci number as the sum of the previous two Fibonacci numbers, so that for $n \geq 3$,

$$F_n = F_{n-1} + F_{n-2}$$

Then $F_3 = 2$, $F_4 = 3$, $F_5 = 5$, $F_6 = 8$, and so on, as in the table below.

n	1	2	3	4	5	6	7	8	9	10	11	12	13
F_n	1	1	2	3	5	8	13	21	34	55	89	144	233

The first 13 Fibonacci numbers

Consequently, the answer to Fibonacci's "immortal rabbits" problem would be $F_{13} = 233$ rabbit pairs (consisting of $F_{12} = 144$ adult pairs and $F_{11} = 89$ baby pairs).

Fibonacci numbers have numerous applications beyond population dynamics, and they appear in nature surprisingly often. For example, the number of petals on a flower is often a Fibonacci number, and the number of spirals on a sunflower, pineapple, and pinecone tends to be a Fibonacci number as well. But what inspires me most about Fibonacci numbers are the beautiful number patterns that they display.

For example, let's look at what happens when we add the first several Fibonacci numbers together:

$$
\begin{aligned}
1 &= 1 &= 2-1 \\
1+1 &= 2 &= 3-1 \\
1+1+2 &= 4 &= 5-1 \\
1+1+2+3 &= 7 &= 8-1 \\
1+1+2+3+5 &= 12 &= 13-1 \\
1+1+2+3+5+8 &= 20 &= 21-1 \\
1+1+2+3+5+8+13 &= 33 &= 34-1 \\
&\;\;\vdots
\end{aligned}
$$

The numbers on the right side of the equation are not quite Fibonacci numbers, but they are close. In fact, each of those numbers is just one shy of being Fibonacci. Let's see why that pattern makes sense. Consider the last equation, and see what happens when we replace each Fibonacci number with the difference of the next two Fibonacci numbers. That is,

$$1+1+2+3+5+8+13$$
$$=(2-1)+(3-2)+(5-3)+(8-5)+(13-8)+(21-13)+(34-21)$$
$$=34-1$$

Notice how the 2 from $(2-1)$ is canceled by the 2 from $(3-2)$. Then the 3 from $(3-2)$ is canceled by the 3 from $(5-3)$. Eventually everything cancels except for the largest term, 34, and the initial -1. In general, this shows that sum of the first n Fibonacci numbers has a simple formula:

$$F_1 + F_2 + F_3 + \cdots + F_n = F_{n+2} - 1$$

Here's a related question with a similarly elegant answer. What do you get when you sum the first n even-positioned Fibonacci numbers? That is, can you simplify the following sum?

$$F_2 + F_4 + F_6 + \cdots + F_{2n}$$

Let's look at some numbers first:

$$
\begin{aligned}
1 &= 1 \\
1+3 &= 4 \\
1+3+8 &= 12 \\
1+3+8+21 &= 33 \\
&\;\;\vdots
\end{aligned}
$$

Wait. These numbers look familiar. In fact, we saw these numbers in our previous sums. They are one less than Fibonacci numbers. In fact, we can transform these numbers into our last problem by using the fact that each Fibonacci number is the sum of two before it, and replacing, after the first term, each even-positioned Fibonacci number with the sum of the two previous Fibonacci numbers, as below.

$$
\begin{aligned}
&\;\; 1 \;\; + \;\; 3 \;\; + \;\; 8 \;\; + \;\; 21 \\
=&\;\; 1 \;\; + \;\; (1+2) \;\; + \;\; (3+5) \;\; + \;\; (8+13) \\
=&\;\; 34 - 1
\end{aligned}
$$

The last line follows from the fact that the sum of the first seven Fibonacci numbers is one less than the ninth.

In general, if we exploit the fact that $F_2 = F_1 = 1$ and replace each subsequent Fibonacci number with the sum of two previous Fibonacci numbers, we see that our sum reduces to the sum of the first $2n - 1$ Fibonacci numbers.

$$
\begin{aligned}
&\;\; F_2 \;\; + \;\; F_4 \;\; + \;\; F_6 \;\; + \cdots + \;\; F_{2n} \\
=&\;\; F_1 \;\; + \;\; (F_2 + F_3) \;\; + \;\; (F_4 + F_5) \;\; + \cdots + (F_{2n-2} + F_{2n-1}) \\
=&\;\; F_{2n+1} \;\; - \;\; 1
\end{aligned}
$$

Let's see what we get when we add the first n odd-positioned Fibonacci numbers.

$$
\begin{aligned}
1 &= 1 \\
1+2 &= 3 \\
1+2+5 &= 8 \\
1+2+5+13 &= 21 \\
&\;\;\vdots
\end{aligned}
$$

Oddly, the pattern is even clearer. The sum of the first n odd-positioned Fibonacci numbers is simply the next Fibonacci number. We can exploit the previous trick, as follows:

$$
\begin{aligned}
& F_1 && + F_3 && + F_5 && + \cdots + && F_{2n-1} \\
={} & 1 && + (F_1 + F_2) && + (F_3 + F_4) && + \cdots + && (F_{2n-3} + F_{2n-2}) \\
={} & 1 && + && (F_{2n} - 1) && \\
={} & F_{2n}
\end{aligned}
$$

> ✂ **Aside**
>
> We could also have arrived at the answer in another way, using what we have already shown. If we subtract the first n even-positioned Fibonacci numbers from the first $2n$ Fibonacci numbers, we'll be left with the first n odd-positioned Fibonacci number:
>
> $$
> \begin{aligned}
> & F_1 + F_3 + F_5 + \cdots + F_{2n-1} \\
> ={} & (F_1 + F_2 + \cdots + F_{2n-1}) - (F_2 + F_4 + \cdots + F_{2n-2}) \\
> ={} & (F_{2n+1} - 1) - (F_{2n-1} - 1) \\
> ={} & F_{2n}
> \end{aligned}
> $$

Counting on Fibonacci

We have only scratched the surface of the beautiful number patterns satisfied by the Fibonacci numbers. You may be tempted to guess that these numbers must be counting something other than pairs of rabbits. Indeed, Fibonacci numbers arise as the solution to many counting problems. In 1150 (before Leonardo of Pisa wrote about rabbits), the Indian poet Hemachandra asked how many cadences of length n were possible if a cadence could contain short syllables of length one or long syllables of length two. We state the question in simpler mathematical terms.

Question: How many ways can we write the number n as a sum of 1s and 2s?

Answer: Let's call the answer f_n and examine f_n for some small values of n.

n	1 – 2 **Sequences That Add to n**	f_n
1	1	1
2	11, 2	2
3	111, 12, 21	3
4	1111, 112, 121, 211, 22	5
5	11111, 1112, 1121, 1211, 122, 2111, 212, 221	8
$\vdots$	$\vdots$	$\vdots$

There is one sum that adds to 1, two sums that add to 2 ($1 + 1$ and 2), and 3 sums that add to 3 ($1 + 1 + 1, 1 + 2, 2 + 1$). Note that we are only allowed to use the numbers 1 and 2 in our sums. Also, the order of the numbers being added matters. So, for example, $1 + 2$ is different from $2 + 1$. There are 5 sums that add to 4 ($1 + 1 + 1 + 1, 1 + 1 + 2, 1 + 2 + 1, 2 + 1 + 1, 2 + 2$). The numbers in our table seem to suggest that the numbers will be Fibonacci numbers, and indeed that is the case.

Let's see why there are $f_5 = 8$ sums that add to the number 5. Such a sum must begin with a 1 or 2. How many of them begin with 1? Well, after the 1, we must have a sequence of 1s and 2s that sum to 4, and we know there are $f_4 = 5$ of them. Likewise, how many of the sums that add to 5 begin with the number 2? After the initial 2, the remaining terms must add to 3, and there are $f_3 = 3$ of those. Hence the total number of sequences that add to 5 must be $5 + 3 = 8$. By the same logic, the number of sequences that add to 6 is 13, since $f_5 = 8$ begin with 1 and $f_4 = 5$ begin with 2. In general, there are f_n sequences that add to n. Of these, f_{n-1} begin with 1 and f_{n-2} begin with 2. Consequently,

$$f_n = f_{n-1} + f_{n-2}$$

Thus, the numbers f_n begin like the Fibonacci numbers and will continue to grow like the Fibonacci numbers. Therefore they *are* the Fibonacci numbers, but with a twist, or perhaps I should say a *shift*. Notice that $f_1 = 1 = F_2$, $f_2 = 2 = F_3$, $f_3 = 3 = F_4$, and so on. (For convenience, we define $f_0 = F_1 = 1$ and $f_{-1} = F_0 = 0$.) In general, we have, for $n \geq 1$,

$$f_n = F_{n+1}$$

Once we know what the Fibonacci numbers count, we can exploit that knowledge to prove many of their beautiful number patterns. Recall the pattern that we saw at the end of Chapter 4 when we summed the diagonals of Pascal's triangle.

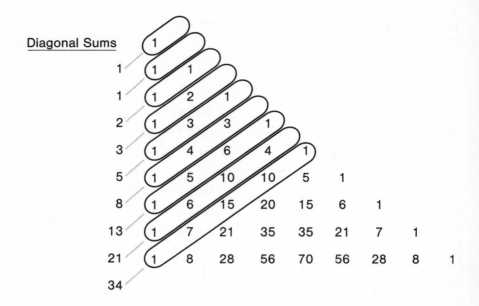

Diagonal Sums

For example, summing the eighth diagonal gave us

$$1 + 7 + 15 + 10 + 1 = 34 = F_9$$

In terms of the "choose numbers," this says

$$\binom{8}{0} + \binom{7}{1} + \binom{6}{2} + \binom{5}{3} + \binom{4}{4} = F_9$$

Let's try to understand this pattern by answering a counting question in two ways.

Question: How many sequences of 1s and 2s add to 8?

Answer 1: By definition, there are $f_8 = F_9$ such sequences.

Answer 2: Let's break this answer into 5 cases, depending on the number of 2s that are used. How many use no 2s? There's just one way to do this, namely 11111111—and not coincidentally, $\binom{8}{0} = 1$.

How many use exactly one 2? This can be done 7 ways: 2111111, 1211111, 1121111, 1112111, 1111211, 1111121, 1111112. Such sequences have 7 numbers, and there are $\binom{7}{1} = 7$ ways to choose the location of the 2.

How many use exactly two 2s? A typical example would be 221111. Instead of listing all 15 of them, note that any such sequence would have two 2s and four 1s, and therefore six digits altogether. There are

$\binom{6}{2} = 15$ ways to choose the locations of the two 2s. By the same logic, a sequence with exactly three 2s would have to have two 1s and therefore have five digits altogether; such sequences could be created in $\binom{5}{3} = 10$ ways. And finally, a sequence with four 2s can only be created $\binom{4}{4} = 1$ way, namely 2222.

Comparing answers 1 and 2 gives us the desired explanation. In general, the same argument can be applied to prove that whenever we sum the nth diagonal of Pascal's triangle, we get a Fibonacci number. Specifically, for all $n \geq 0$, when we sum the nth diagonal (until we fall off the triangle after about $n/2$ terms), we get

$$\binom{n}{0} + \binom{n-1}{1} + \binom{n-2}{2} + \binom{n-3}{3} + \cdots = f_n = F_{n+1}$$

An equivalent and more visual way to think about the Fibonacci numbers is through *tilings*. For example, $f_4 = 5$ counts the five ways to tile a strip of length 4 using squares (of length 1) and dominos (of length 2). For example, the sum $1 + 1 + 2$ is represented by the tiling square-square-domino.

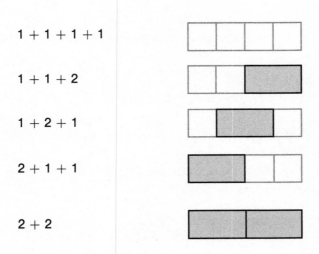

$$1 + 1 + 1 + 1$$

$$1 + 1 + 2$$

$$1 + 2 + 1$$

$$2 + 1 + 1$$

$$2 + 2$$

There are 5 tilings of length 4, using squares and dominos, illustrating $f_4 = 5$

We can use tilings to understand another remarkable Fibonacci number pattern. Let's look at what happens when we square the Fibonacci numbers.

n	0	1	2	3	4	5	6	7	8	9	10
f_n	1	1	2	3	5	8	13	21	34	55	89
f_n^2	1	1	4	9	25	64	169	441	1156	3025	7921

The squares of Fibonacci numbers, from f_0 to f_{10}

Now, it's no surprise that when you add consecutive Fibonacci numbers, you get the next Fibonacci number. (That's how they're created, after all.) But you wouldn't expect anything interesting to happen with the squares. However, check out what happens when we add consecutive squares together:

$$
\begin{aligned}
f_0^2 + f_1^2 &= 1^2 + 1^2 &= 1 + 1 &= 2 &= f_2 \\
f_1^2 + f_2^2 &= 1^2 + 2^2 &= 1 + 4 &= 5 &= f_4 \\
f_2^2 + f_3^2 &= 2^2 + 3^2 &= 4 + 9 &= 13 &= f_6 \\
f_3^2 + f_4^2 &= 3^2 + 5^2 &= 9 + 25 &= 34 &= f_8 \\
f_4^2 + f_5^2 &= 5^2 + 8^2 &= 25 + 64 &= 89 &= f_{10} \\
&\vdots
\end{aligned}
$$

Let's try to explain this pattern in terms of counting. The last equation says that

$$f_4^2 + f_5^2 = f_{10}$$

Why should that be? We can explain this by asking a simple counting question.

Question: How many ways can you tile a strip of length 10 using squares and dominos?

Answer 1: By definition, there are f_{10} such tilings. Here is a typical tiling, which represents the sum $2 + 1 + 1 + 2 + 1 + 2 + 1$.

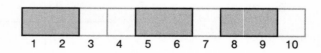

We say that this tiling is *breakable* at cells 2, 3, 4, 6, 7, 9, and 10. (Equivalently, the tiling is breakable everywhere except the middle of a domino. In this example, it is *unbreakable* at cells 1, 5, and 8.)

Answer 2: Let's break the answer into two cases: those tilings that are breakable at cell 5, and those tilings that are not. How many ways can we create a length-10 tiling that is breakable at cell 5? Such a tiling can be split into two halves, where the left half can be tiled in $f_5 = 8$ ways and the right half can also be tiled in $f_5 = 8$ ways. Hence by the rule of product in Chapter 4, we can create such a sum in $f_5^2 = 8^2$ ways, as illustrated below.

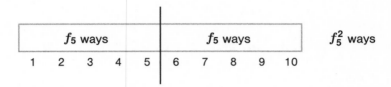

There are f_5^2 tilings of length 10 that are breakable at cell 5

How many length-10 tilings are *not* breakable at cell 5? Such tilings must necessarily have a domino covering cells 5 and 6, as illustrated below. Now the left half and right half can each be tiled $f_4 = 5$ ways and so there are $f_4^2 = 5^2$ unbreakable tilings. Putting these two cases together, it follows that $f_{10} = f_5^2 + f_4^2$, as desired.

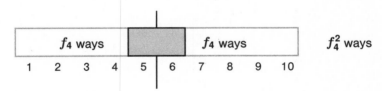

There are f_4^2 tilings of length 10 that are not breakable at cell 5

In general, by considering whether a tiling of length $2n$ is breakable in the middle or not, we arrive at the beautiful pattern

$$f_{2n} = f_n^2 + f_{n-1}^2$$

> ⋈ **Aside**
>
> After seeing the previous identity, we might try to extend it to similar cases. For example, consider the number of tilings of length $m + n$. How many of these tilings are breakable at cell m? The left side can be tiled f_m ways and the right side can be tiled f_n ways, so there are $f_m f_n$ such tilings. How many are *not* breakable at m? Such a tiling must have a domino covering cells m and $m + 1$, and the rest of the tiling can be tiled $f_{m-1} f_{n-1}$ ways. Altogether, we get the following useful identity. For $m, n \geq 0$,
>
> $$f_{m+n} = f_m f_n + f_{m-1} f_{n-1}$$

Time for a new pattern. Let's see what happens when we add all of the squares of Fibonacci numbers together.

$$
\begin{aligned}
1^2 + 1^2 &= & 2 &= 1 \times 2 \\
1^2 + 1^2 + 2^2 &= & 6 &= 2 \times 3 \\
1^2 + 1^2 + 2^2 + 3^2 &= & 15 &= 3 \times 5 \\
1^2 + 1^2 + 2^2 + 3^2 + 5^2 &= & 40 &= 5 \times 8 \\
1^2 + 1^2 + 2^2 + 3^2 + 5^2 + 8^2 &= & 104 &= 8 \times 13 \\
&\vdots&
\end{aligned}
$$

Wow, this is so cool! The sum of the squares of Fibonacci numbers is the product of the last two! But why should the sum of the squares of 1, 1, 2, 3, 5, and 8 add to 8×13? One way to "see" this with geometric figures is to take six squares with side lengths 1, 1, 2, 3, 5, and 8 and assemble them as in the figure below.

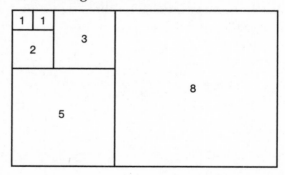

Start with a 1-by-1 square, then put the other 1-by-1 square next to it, creating a 1-by-2 rectangle. Beneath that rectangle, put the 2-by-2 square, which creates a 3-by-2 rectangle. Next to the long edge of that rectangle, place the 3-by-3 square (creating a 3-by-5 rectangle), then place the 5-by-5 square underneath that (creating an 8-by-5 rectangle). Finally, place the 8-by-8 square next to that, creating one giant 8-by-13 rectangle. Now let's ask a simple question.

Question: What is the area of the giant rectangle?

Answer 1: On the one hand, the area of the rectangle is the sum of the areas of the squares that compose it. In other words, the area of this rectangle must be $1^2 + 1^2 + 2^2 + 3^2 + 5^2 + 8^2$.

Answer 2: On the other hand, we have a big rectangle with height 8 and a base length of $5 + 8 = 13$, hence the area of this rectangle must be 8×13.

Since both answers are correct, they must give the same area, which explains the last identity. In fact, if you read again how the rectangle was constructed, you see that it explains all of the relationships listed about this pattern (like $1^2 + 1^2 + 2^2 + 3^2 + 5^2 = 5 \times 8$). And if you continue this logic, you will create rectangles of size 13×21, 21×34, and so on, so the pattern will continue forever. The general formula says that

$$1^2 + 1^2 + 2^2 + 3^2 + 5^2 + 8^2 + \cdots + F_n^2 = F_n F_{n+1}$$

Now let's see what happens when we multiply Fibonacci *neighbors* together. For example, the neighbors of 5 are 3 and 8, and their product is $3 \times 8 = 24$, which is one less than 5^2. The neighbors of 8 are 5 and 13, and their product is $5 \times 13 = 65$, which is just one more than 8^2. Examining the table below, it's hard to resist the conclusion that the product of the neighbors is always one away from the square of the original Fibonacci number. In other words,

$$F_n^2 - F_{n-1} F_{n+1} = \pm 1$$

n	1	2	3	4	5	6	7	8	9	10	11
F_n	1	1	2	3	5	8	13	21	34	55	89
F_n^2	1	1	4	9	25	64	169	441	1156	3025	7921
$F_{n-1} F_{n+1}$	0	2	3	10	24	65	168	442	1155	3026	7920
$F_n^2 - F_{n-1} F_{n+1}$	1	−1	1	−1	1	−1	1	−1	1	−1	1

The product of the neighbors of a Fibonacci number is one away from its square

Using a proof technique (called induction) that we will learn in the next chapter, it can be proved that for $n \geq 1$,

$$F_n^2 - F_{n-1} F_{n+1} = (-1)^{n+1}$$

Let's push this pattern even further by looking at *distant* neighbors. Look at the Fibonacci number $F_5 = 5$. We saw that when we multiply its immediate neighbors, we get $3 \times 8 = 24$, which is one away from 5^2. But the same thing happens when we multiply the Fibonacci numbers that are two away from it: $2 \times 13 = 26$, which is also one away from 5^2. How about the neighbors that are 3 away, 4 away, or 5 away? Their products are $1 \times 21 = 21$, $1 \times 34 = 34$, and $0 \times 55 = 0$. How far away are these numbers from 25? They are 4 away, 9 away, and 25 away, which are all perfect squares. But they're not just *any* perfect squares; they are squares of Fibonacci numbers! See the table below for more evidence of this pattern. The general pattern is:

$$F_n^2 - F_{n-r}F_{n+r} = \pm F_r^2$$

n	1	2	3	4	5	6	7	8	9	10	
F_n	1	1	2	3	5	8	13	21	34	55	
F_n^2	1	1	4	9	25	64	169	441	1156	3025	$F_n^2 - F_{n-r}F_{n+r}$
$F_{n-1}F_{n+1}$	0	2	3	10	24	65	168	442	1155	3026	± 1
$F_{n-2}F_{n+2}$		0	5	8	26	63	170	440	1157	3024	± 1
$F_{n-3}F_{n+3}$			0	13	21	68	165	445	1152	3029	± 4
$F_{n-4}F_{n+4}$				0	34	55	178	432	1165	3016	± 9
$F_{n-5}F_{n+5}$					0	89	144	466	1131	3050	± 25
$\vdots$					$\vdots$						$\vdots$

The product of distant neighbors of a Fibonacci number is always close to its square. Their distance apart is always the square of a Fibonacci number.

More Fibonacci Patterns

In Pascal's triangle, we saw that the even and odd numbers displayed a startlingly complex pattern. With Fibonacci numbers, the situation is much simpler. Which Fibonacci numbers are even?

$$1, 1, 2, 3, 5, 8, 13, 21, 34, 55, 89, 144 \ldots$$

The even numbers are $F_3 = 2$, $F_6 = 8$, $F_9 = 34$, $F_{12} = 144$, and so on. (In this section, we switch back to the capital-F Fibonacci numbers, since they will produce prettier patterns.) The even numbers first appear in positions $3, 6, 9, 12$, which suggests that they occur precisely every 3 terms. We can prove this by noticing that the pattern begins

odd, odd, even

and is then forced to repeat itself,

odd, odd, even, odd, odd, even, odd, odd, even ...

since after any "odd, odd, even" block, the next block must begin with "odd + even = odd," then "even + odd = odd," followed by "odd + odd = even," so the pattern goes on forever.

In the language of congruences from Chapter 3, every even number is congruent to 0 (mod 2), every odd number is congruent to 1 (mod 2), and $1 + 1 \equiv 0$ (mod 2). Thus the mod 2 version of the Fibonacci numbers looks like

$$1, 1, 0, 1, 1, 0, 1, 1, 0, 1, 1, 0 \ldots$$

What about multiples of 3? The first multiples of 3 are $F_4 = 3, F_8 = 21, F_{12} = 144$, so it's tempting to believe that the multiples of 3 occur at every fourth Fibonacci number. To verify this, let's reduce the Fibonacci numbers to 0, 1, or 2 and use mod 3 arithmetic, where

$$1 + 2 \equiv 0 \text{ and } 2 + 2 \equiv 1 \quad (\text{mod } 3)$$

The mod 3 version of the Fibonacci numbers is

$$1, 1, 2, 0, 2, 2, 1, 0, \quad 1, 1, 2, 0, 2, 2, 1, 0, \quad 1, 1 \ldots$$

After eight terms, we are back to the beginning with 1 and 1, so the pattern will repeat in blocks of size eight, with 0 in every fourth position. Thus every fourth Fibonacci number is a multiple of 3, and vice versa. Using arithmetic mod 5 or 8 or 13, you can verify that

> *Every fifth Fibonacci number is a multiple of 5*
> *Every sixth Fibonacci number is a multiple of 8*
> *Every seventh Fibonacci number is a multiple of 13*

and the pattern continues.

What about *consecutive* Fibonacci numbers? Do they have anything in common? What's interesting here is that we can show that, in one sense, they have *nothing* in common. We say that the consecutive pairs of Fibonacci numbers

$$(1, 1), (1, 2), (2, 3), (3, 5), (5, 8), (8, 13), (13, 21), (21, 34), \ldots$$

are *relatively prime*, which means that there is no number bigger than 1 that divides both numbers. For example, if we look at the last pair, we see that 21 is divisible by 1, 3, 7, and 21. The divisors of 34 are 1, 2, 17, and 34. Thus 21 and 34 have no common divisors, except for 1. How can we be sure that this pattern will continue forever? How do we know that the next pair $(34, 55)$ is guaranteed to be relatively prime? We can prove this without finding the factors of 55. Suppose, to the contrary, that there was a number $d > 1$ that divided both 34 and 55. But then such a number would have to divide their difference $55 - 34 = 21$ (since if 55 and 34 are multiples of d then so is their difference). However, this is impossible, since we already know that there is no number $d > 1$ that divides both 21 and 34. Repeating this argument guarantees that all consecutive pairs of Fibonacci numbers will be relatively prime.

And now for my favorite Fibonacci fact! The *greatest common divisor* of two numbers is the largest number that divides both of the numbers. For example, the greatest common divisor of 20 and 90 is 10, denoted

$$\gcd(20, 90) = 10$$

What do you think is the greatest common divisor of the 20th Fibonacci number and the 90th Fibonacci number? The answer is absolutely poetic. The answer is 55, which is also a Fibonacci number. Moreover, it happens to be the 10th Fibonacci number! In equations,

$$\gcd(F_{20}, F_{90}) = F_{10}$$

And in general, we have for integers m and n,

$$\gcd(F_m, F_n) = F_{\gcd(m,n)}$$

In other words, "the gcd of the Fs is the F of the gcd"! We won't prove this beautiful fact here, but I couldn't resist showing it to you.

Sometimes a pattern can be deceptive. For instance, which of the Fibonacci numbers are prime numbers? (As we'll discuss in the next chapter, a prime is a number bigger than 1 that is divisible only by 1 and itself.) Numbers bigger than 1 that are not prime are called *composite* since they can be de*composed* into the product of smaller numbers. The first few prime numbers are

$$2, 3, 5, 7, 11, 13, 17, 19 \ldots$$

Now look at the Fibonacci numbers that are in prime locations:

$$F_2 = 1, F_3 = 2, F_5 = 5, F_7 = 13, F_{11} = 89, F_{13} = 233, F_{17} = 1597$$

The numbers $2, 5, 13, 89, 233$, and 1597 are prime. The pattern suggests that if $p > 2$ is prime, then so is F_p, but the pattern fails with the next data point. $F_{19} = 4181$ is not prime, since $4181 = 37 \times 113$. However, it *is* true that every prime Fibonacci number bigger than 3 will occur in a prime position. This follows from our earlier pattern. F_{14} must be composite, since every 7th Fibonacci number is a multiple of $F_7 = 13$ (and indeed, $F_{14} = 377 = 13 \times 29$).

In fact, Fibonacci primes seem to be pretty rare. As of this writing, there have only been 33 confirmed discoveries of prime Fibonacci numbers, the largest one being F_{81839}, and it is still an open question in mathematics whether there is an infinite number of Fibonacci primes.

We interrupt this serious discussion to bring you a fun little magic trick based on Fibonacci numbers.

Row 1:	3
Row 2:	7
Row 3:	10
Row 4:	17
Row 5:	27
Row 6:	44
Row 7:	71
Row 8:	115
Row 9:	186
Row 10:	301

A Fibonacci magic trick: Start with any two positive numbers in rows 1 and 2. Fill out the rest of the table in Fibonacci fashion $(3 + 7 = 10, 7 + 10 = 17$, and so on), then divide row 10 by row 9. The answer is guaranteed to begin with 1.61.

On the table above, write any numbers between 1 and 10 in the first and second row. Add those numbers together and put the total in row 3. Add the numbers in row 2 and row 3, and put their total in row 4. Continue filling out the table in Fibonacci fashion (row 3 + row 4 = row 5, and so on) until you have numbers in all 10 rows. Now divide the number in row 10 by the number in row 9 and read off the first three digits, including the decimal point. In this example, we see $\frac{301}{186} =$ 1.618279... so its first three digits are 1.61. Believe it or not, starting with *any* two positive numbers in rows 1 and 2 (they don't have to be whole numbers or be between 1 and 10), the ratio of row 10 and row 9 is guaranteed to be 1.61. Try an example or two yourself.

To see why this trick works, let's let x and y denote the numbers in rows 1 and 2, respectively. Then by the Fibonacci rules, row 3 must be $x + y$, row 4 must be $y + (x + y) = x + 2y$, and so on, as illustrated in the table below.

Row 1:	x
Row 2:	y
Row 3:	$x + y$
Row 4:	$x + 2y$
Row 5:	$2x + 3y$
Row 6:	$3x + 5y$
Row 7:	$5x + 8y$
Row 8:	$8x + 13y$
Row 9:	$13x + 21y$
Row 10:	$21x + 34y$

The question amounts to looking at the ratio of the entries in rows 10 and 9:

$$\frac{\text{Row 10}}{\text{Row 9}} = \frac{21x + 34y}{13x + 21y}$$

So why should that ratio always begin with 1.61? The answer is based on the idea of adding fractions *incorrectly*. Suppose you have two fractions $\frac{a}{b}$ and $\frac{c}{d}$ where b and d are positive. What happens when you add the numerators together and add the denominators together? Believe it or not, the resulting number, called the *mediant*, will always be somewhere in between the original two numbers. That is, for any distinct fractions $a/b < c/d$, where b and d are positive, we have

$$\frac{a}{b} < \frac{a + c}{b + d} < \frac{c}{d}$$

For example, starting with the fractions $1/3$ and $1/2$, their mediant is $2/5$, which lies between them: $1/3 < 2/5 < 1/2$.

Now notice that for $x, y > 0$,

$$\frac{21x}{13x} = \frac{21}{13} = 1.615\ldots$$

$$\frac{34y}{21y} = \frac{34}{21} = 1.619\ldots$$

Hence their mediant must lie in between them. In other words,

$$1.615\ldots = \frac{21}{13} = \frac{21x}{13x} < \frac{21x + 34y}{13x + 21y} < \frac{34y}{21y} = \frac{34}{21} = 1.619\ldots$$

Thus the ratio of row 10 and row 9 must begin with the digits 1.61, as predicted!

What is the significance of the number 1.61? If you were to extend the table further and further, you would find that the ratio of consecutive terms will gradually get closer and closer to the golden ratio

$$g = \frac{1 + \sqrt{5}}{2} = 1.61803\ldots$$

Mathematicians sometimes denote this number with the Greek letter ϕ, spelled phi, and pronounced "phie" (rhymes with pie) or "fee" as in "phi-bonacci."

> ✂ **Aside**
>
> Using algebra, we can prove that the ratio of consecutive Fibonacci numbers gets closer and closer to g. Suppose that F_{n+1}/F_n gets closer and closer to some ratio r as n gets larger and larger. But by the definition of Fibonacci numbers, $F_{n+1} = F_n + F_{n-1}$, so
>
> $$\frac{F_{n+1}}{F_n} = \frac{F_n + F_{n-1}}{F_n} = 1 + \frac{F_{n-1}}{F_n}$$
>
> Now as n gets larger and larger, the left side approaches r and the right side approaches $1 + \frac{1}{r}$. Thus,
>
> $$r = 1 + \frac{1}{r}$$
>
> When we multiply both sides of this equation by r, we get
>
> $$r^2 = r + 1$$
>
> In other words, $r^2 - r - 1 = 0$, and by the quadratic formula, the only positive solution is $r = \frac{1+\sqrt{5}}{2}$, which is g.

There's an intriguing formula for the nth Fibonacci number that uses g, called **Binet's formula.** It says that

$$F_n = \frac{1}{\sqrt{5}}\left[\left(\frac{1+\sqrt{5}}{2}\right)^n - \left(\frac{1-\sqrt{5}}{2}\right)^n\right]$$

I find it amazing and amusing that this formula, with all of the $\sqrt{5}$ terms running around, produces whole numbers!

We can simplify Binet's formula somewhat, since the quantity

$$\frac{1-\sqrt{5}}{2} = -0.61803\ldots \text{ (same dots as before!)}$$

is between -1 and 0, and when we raise it to higher and higher powers, it gets closer and closer to 0. In fact, it can be shown that for any $n \geq 0$, you can calculate F_n by taking $g^n/\sqrt{5}$ and rounding it to the nearest integer. Go ahead and take out your calculator and try this. If you approximate g with 1.618, then raise 1.618 to the 10th power, you get $122.966\ldots$ (which is suspiciously close to 123). Then divide that number by $\sqrt{5} \approx 2.236$, and you get 54.992. Rounding this number tells us that $F_{10} = 55$, which we already knew. If we take g^{20} we get 15126.99993, and if we divide it by $\sqrt{5}$ we get 6765.00003, so $F_{20} = 6765$. Using our calculator to compute $g^{100}/\sqrt{5}$ tells us that F_{100} is about 3.54×10^{20}.

In the calculations we just did, it seemed that g^{10} and g^{20} were also practically whole numbers. What's going on there? Behold the Lucas (pronounced "loo-kah") numbers,

$$1, 3, 4, 7, 11, 18, 29, 47, 76, 123, 199, 322, 521\ldots$$

named in honor of French mathematician Édouard Lucas (1842–1891) who discovered many beautiful properties of these numbers and Fibonacci numbers, including the greatest common divisor property we saw earlier. Indeed, Lucas was the first person to name the sequence $1, 1, 2, 3, 5, 8 \ldots$ the Fibonacci numbers. The Lucas numbers satisfy their own (somewhat simpler) version of Binet's formula, namely

$$L_n = \left(\frac{1 + \sqrt{5}}{2} \right)^n + \left(\frac{1 - \sqrt{5}}{2} \right)^n$$

Put another way, for $n \geq 1$, L_n is the integer closest to g^n. (This is consistent with what we saw earlier since $g^{10} \approx 123 = L_{10}$.) We can see more connections with Fibonacci and Lucas numbers in the table below.

n	1	2	3	4	5	6	7	8	9	10
F_n	1	1	2	3	5	8	13	21	34	55
L_n	1	3	4	7	11	18	29	47	76	123
$F_{n-1} + F_{n+1}$		3	4	7	11	18	29	47	76	123
$L_{n-1} + L_{n+1}$		5	10	15	25	40	65	105	170	275
$F_n L_n$	1	3	8	21	55	144	377	987	2584	6765

Fibonacci numbers, Lucas numbers, and some of their interactions

It's hard to miss some of the patterns. For example, when we add Fibonacci neighbors, we get Lucas numbers:

$$F_{n-1} + F_{n+1} = L_n$$

and when we add Lucas neighbors we get 5 times the corresponding Fibonacci number,

$$L_{n-1} + L_{n+1} = 5F_n$$

When we multiply corresponding Fibonacci and Lucas numbers together, we get another Fibonacci number!

$$F_n L_n = F_{2n}$$

> **⋈ Aside**
>
> Let's prove this last relationship using the Binet formulas and a little bit of algebra (specifically $(x - y)(x + y) = x^2 - y^2$). Letting $h = (1 - \sqrt{5})/2$, the Binet formulas for Fibonacci and Lucas numbers can be stated as
>
> $$F_n = \frac{1}{\sqrt{5}}(g^n - h^n) \text{ and } L_n = g^n + h^n$$
>
> When we multiply these expressions together, we get
>
> $$F_n L_n = \frac{1}{\sqrt{5}}(g^n - h^n)(g^n + h^n) = \frac{1}{\sqrt{5}}\left(g^{2n} - h^{2n}\right) = F_{2n}$$

So where does the name "golden ratio" come from? It comes from the golden rectangle below, where the ratio of the long side to the short side is exactly $g = 1.61803\ldots$.

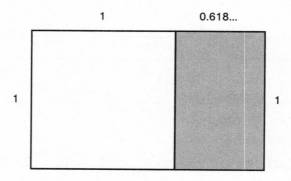

The golden rectangle produces a smaller rectangle that also has golden proportions

If we label the short side as 1 unit and remove a 1-by-1 square from the rectangle, then the resulting rectangle has dimensions 1-by-$(g - 1)$ and the ratio of its long side to short side is

$$\frac{1}{g - 1} = \frac{1}{0.61803\ldots} = 1.61803\ldots = g$$

Thus the smaller rectangle has the same proportions as the original rectangle. By the way, g is the only number that has this nice property, since the equation $\frac{1}{g-1} = g$ implies that $g^2 - g - 1 = 0$, and by the quadratic formula, the only positive number that satisfies this is $(1 + \sqrt{5})/2 = g$.

Because of this property, the golden rectangle is considered by some to be the most aesthetically beautiful rectangle, and it has been used deliberately by various artists, architects, and photographers in their work. Luca Pacioli, a longtime friend and collaborator of Leonardo da Vinci, gave the golden rectangle the name "the divine proportion."

The Fibonacci numbers and the golden ratio have inspired many artists, architects, and photographers. (Photo courtesy of Natalya St. Clair)

Because there are so many marvelous mathematical properties satisfied by the golden ratio, there is sometimes a tendency to see it even in places where it doesn't exist. For example, in the book *The Da Vinci Code*, author Dan Brown asserts that the number 1.618 appears everywhere, and that the human body is practically a testament to that number. For instance, it is claimed that the ratio of a person's height to the height of his or her bellybutton is always 1.618. Now, I personally have not conducted this experiment, but according to the *College Mathematics Journal* article "Misconceptions About the Golden Ratio," by George Markowski, this is simply not true. But to some people, anytime a number appears to be close to 1.6, they assume it must be the golden ratio at work.

I have often said that many of the number patterns satisfied by Fibonacci numbers are pure poetry in motion. Well, here's an example of where the Fibonacci numbers actually arise in poetry. Most limericks have the following meter. (You might call it a "dum" limerick!)

Limerick	di	dum	Syllables
di-dum di-di-dum di-di-dum	5	3	8
di-dum di-di-dum di-di-dum	5	3	8
di-dum di-di-dum	3	2	5
di-dum di-di-dum	3	2	5
di-dum di-di-dum di-di-dum	5	3	8
Total	**21**	**13**	**34**

The poetry of Fibonacci numbers

If you count the syllables in each row, we see Fibonacci numbers everywhere! Inspired by this, I decided to write a Fibonacci limerick of my own.

I think Fibonacci is fun.
It starts with a 1 and a 1.
Then 2, 3, 5, 8.
But don't stop there, mate.
The fun has just barely begun!

CHAPTER SIX

$1 + 2 + 3 = 1 \times 2 \times 3 = 6$

The Magic of Proofs

The Value of Proofs

One of the great joys of doing mathematics, and indeed what separates mathematics from all other sciences, is the ability to prove things true beyond a shadow of a doubt. In other sciences, we accept certain laws because they conform to the real world, but those laws can be contradicted or modified if new evidence presents itself. But in mathematics, if a statement is proved to be true, then it is true forever. For example, it was proved by Euclid over two thousand years ago that there are infinitely many prime numbers, and there is nothing that we can say or do that will ever contradict the truth of that statement. Technology comes and goes, but a theorem is forever. As the great mathematician G. H. Hardy said, "A mathematician, like a painter or poet, is a maker of patterns. If his patterns are more permanent than theirs, it is because they are made with ideas." To me, it often seems like the best route to academic immortality would be to prove a new mathematical theorem.

Not only can mathematics prove things with absolute certainty, it can also prove that some things are *impossible*. Sometimes people say, "You can't prove a negative," which I suppose means that you can't prove that there are no purple cows, since one might show up someday. But in mathematics, you *can* prove negatives. For example, no matter how hard you try, you will never be able to find two even numbers that add up to an odd number nor find a prime number that is larger than all other primes. Proofs can be a little scary the first (or second or third) time that you encounter them, and they are definitely an acquired taste. But once you get the hang of them, proofs can be quite fun to read and write. A good proof is like a well-told joke or story—it leaves you feeling very satisfied at the end.

Let me tell you about one of my first experiences of proving something impossible. When I was a kid, I loved games and puzzles. One day, a friend challenged me with a puzzle about a game, and so of course I was interested. He showed me an empty 8-by-8 checkerboard and brought out 32 regular 1-by-2 dominos. He asked, "Can you cover the checkerboard using all 32 dominos?" I said, "Of course, just lay 4 dominos across each row, like this."

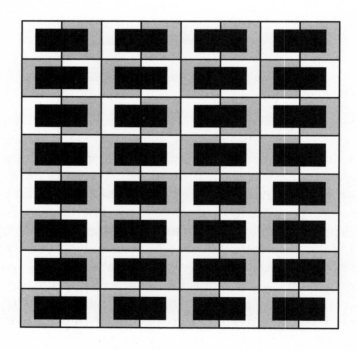

Covering an 8-by-8 checkerboard with dominos

"Very good," he said. "Now suppose I remove the squares from the lower-right and upper-left corners." He placed a coin on those two squares so I couldn't use them. Now can you cover the remaining 62 squares using 31 dominos?"

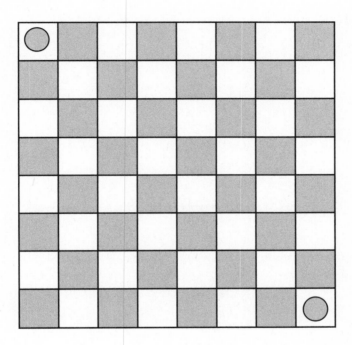

Can the checkerboard be covered with dominos when two opposite corners are removed?

"Probably," I said. But no matter how hard I tried, I was unable to do it. I was starting to think that the task was impossible.

"If you think it's impossible, can you *prove* that it's impossible?" my friend asked. But how could I prove that it was impossible without exhaustively going through the zillions of possible failed attempts? He then suggested, "Look at the colors on the checkerboard."

The colors? What did that have to do with anything? But then I saw it. Since both of the removed corner squares were light-colored, then the remaining board consisted of 32 dark squares and just 30 light squares. And since every domino covers exactly one light square and one dark square, it would be impossible for 31 dominos to tile such a board. Cool!

✕ Aside

If you liked the last proof, you will enjoy this one as well. The video game Tetris uses seven pieces of different shapes, sometimes given the names I, J, L, O, Z, T, and S.

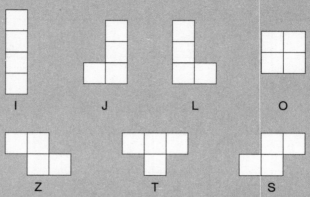

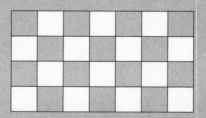

Can these 7 pieces be arranged to form a 4-by-7 rectangle?

Each piece occupies exactly four squares, so it is natural to wonder whether it would be possible to arrange them in such a way that they form a 4-by-7 rectangle, where we are allowed to flip or rotate the tiles. This turns out to be an impossible challenge. How do you prove that it's impossible? Color the rectangle with fourteen light squares and fourteen dark squares, as shown.

Notice that, with the exception of the T tile, every piece must cover two light squares and two dark squares, no matter where it is located. But the T tile must cover three squares of one color and one square of the other color. Consequently, no matter where the other six tiles are located, they must cover exactly twelve light squares and twelve dark squares. This leaves two light squares and two dark squares to be covered by the T tile, which is impossible.

So if we have a mathematical statement that we think is true, how do we *actually* go about convincing ourselves? Typically, we start by describing the mathematical objects that we are working with, like the collection of *integers*

$$\ldots, -2, -1, 0, 1, 2, 3, \ldots$$

consisting of all whole numbers: positive numbers, negative numbers, and zero.

Once we have described our objects, we make assumptions about these objects that we consider to be self-evident, like "The sum or product of two integers is always an integer." (In the next chapter, on geometry, we'll assume statements like "For any two points there is a line that goes through those two points.") These self-evident statements are called *axioms*. And from these axioms, and a little logic and algebra, we can often derive other true statements (called *theorems*) that are sometimes far from obvious. In this chapter, you will learn the basic tools of proving mathematical statements.

Let's start by proving some theorems that are easy to believe. The first time we hear a statement like "The sum of two even numbers is even" or "The product of two odd numbers is odd," our mind typically checks the statement with a few examples, then concludes that it's probably true or that it makes sense. You might even think that the statement is *so* self-evident that we could make it an axiom. But we don't need to do that, since we can *prove* this statement using the axioms we already know. To prove something about even and odd numbers, we need to be clear about what odd and even mean.

An *even number* is a multiple of 2. To express this algebraically, we say n is even if $n = 2k$ where k is an integer. Is 0 an even number? Yes, since $0 = 2 \times 0$. We are now ready to prove that the sum of two even numbers is even.

Theorem: If m and n are even numbers, then $m + n$ is also an even number.

This is an example of an "if-then" theorem. To prove such a statement, we typically assume the "if" part and, through a mixture of logic and algebra, show that the "then" part follows from our assumptions. Here we assume that m and n are even and want to conclude that $m + n$ is also even.

Proof: Suppose m and n are even numbers. Thus, $m = 2j$ and $n = 2k$ where j and k are integers. But then

$$m + n = 2j + 2k = 2(j + k)$$

and since $j + k$ is an integer, $m + n$ is also a multiple of 2, so $m + n$ must be even.   □

Notice that the proof relied on the axiom that the sum of two integers (here, $j + k$) must also be an integer. As you prove more complicated statements, you are more likely to rely on previously proved theorems than the basic axioms. Mathematicians often denote the end of a proof with a designation like □ or ■ or Q.E.D on the right margin of the last line of the proof, as seen above. Q.E.D. is the abbreviation for the Latin phrase "quod erat demonstrandum," meaning "that which was to be demonstrated." (It's also the abbreviation for the English phrase "quite easily done," if you prefer.) If I think that a proof is particularly elegant, I will end it with a smiley like ☺.

After proving an if-then theorem, mathematicians can't resist wondering about the truth of the *converse* statement, obtained by reversing the "if" part and the "then" part. Here, the converse statement would be "If $m + n$ is an even number, then m and n are even numbers." It's easy to see that this statement is false, simply by providing a *counterexample*. For this statement, the counterexample is literally as easy as

$$1 + 1 = 2$$

since this example shows that it's possible to have an even sum even when both numbers are not even.

Our next theorem is about *odd numbers*. An odd number is a number that is *not* a multiple of 2. Consequently, when you divide an odd number by 2, you always get a remainder of 1. Algebraically, n is odd if $n = 2k + 1$, where k is an integer. This allows us to prove, by simple algebra, that the product of odd numbers is odd.

Theorem: If m and n are odd, then mn is odd.

Proof: Suppose m and n are odd numbers. Then $m = 2j + 1$ and $n = 2k + 1$ for some integers j and k. Thus, according to FOIL,

$$mn = (2j + 1)(2k + 1) = 4jk + 2j + 2k + 1 = 2(2jk + j + k) + 1$$

and since $2jk + j + k$ is an integer, this means that the number mn is of the form "twice an integer + 1." Hence mn is odd.   □

What about the converse statement: "If mn is odd, then m and n are odd"? This statement is also true, and we can prove it using a *proof by contradiction*. In a proof by contradiction, we show that rejecting the conclusion (here, the conclusion is that m and n are odd) leads to a problem. In particular, if we reject the conclusion, there would also be a problem with our assumptions, so logically the conclusion must be true.

Theorem: If mn is odd, then m and n are odd.

Proof: Suppose, to the contrary, that m or n (or both) were even. It really doesn't matter which one is even, so let's say that m is even and therefore $m = 2j$ for some integer j. Then the product $mn = 2jn$ is also even, contradicting our initial assumption that mn is odd. □

When a statement and its converse are both true, mathematicians often call this an "if and only if theorem." We have just shown the following:

Theorem: m and n are odd if and only if mn is odd.

Rational and Irrational Numbers

The theorems just presented were probably not surprising to you and their proofs were pretty straightforward. The fun begins when we prove theorems that are less intuitive. So far, we have mostly been dealing with integers, but now it's time to graduate to fractions. The *rational* numbers are those numbers that can be expressed as a fraction. To be more precise, we say r is rational if $r = a/b$, where a and b are integers (and $b \neq 0$). Rational numbers include, for example, $23/58$, $-22/7$, and 42 (which equals $42/1$). Numbers that are not rational are called *irrational*. (You may have heard that the number $\pi = 3.14159\ldots$ is irrational, and we will have more to say about that in Chapter 8.)

For the next theorem, it may be helpful to recall how to add fractions. It is easiest to add fractions when they have a common denominator. For example,

$$\frac{1}{5} + \frac{2}{5} = \frac{3}{5}, \quad \frac{3}{4} + \frac{1}{4} = \frac{4}{4} = 1, \quad \frac{5}{8} + \frac{7}{8} = \frac{12}{8} = \frac{3}{2} = 1.5$$

Otherwise, to add the fractions, we rewrite them to have the same denominators. For example,

$$\frac{1}{3} + \frac{1}{6} = \frac{2}{6} + \frac{1}{6} = \frac{3}{6} = \frac{1}{2}, \quad \frac{2}{7} + \frac{3}{5} = \frac{10}{35} + \frac{21}{35} = \frac{31}{35}$$

In general, we can add any two fractions a/b and c/d by giving them a common denominator, as follows:

$$\frac{a}{b} + \frac{c}{d} = \frac{ad}{bd} + \frac{bc}{bd} = \frac{ad + bc}{bd}$$

We are now ready to prove a simple fact about rational numbers.

Theorem: The average of two rational numbers is also rational.

Proof: Let x and y be rational numbers. Then there exist integers a, b, c, d where $x = a/b$ and $y = c/d$. Notice x and y have average

$$\frac{x+y}{2} = \frac{a/b + c/d}{2} = \frac{ad + bc}{2bd}$$

and that average is a fraction where the numerator and denominator are integers. Consequently, the average of x and y is rational.

Let's think about what this theorem is saying. It says that for any two rational numbers, even if they are really, really close together, we can always find another rational number in between them. You might be tempted to conclude that all numbers are rational (as the ancient Greeks believed for a while). But, surprisingly, that's not the case. Let's consider the number $\sqrt{2}$, which has decimal expansion $1.4142\ldots$. Now, there are many ways to *approximate* $\sqrt{2}$ with a fraction. For example, $\sqrt{2}$ is approximately $10/7$ or $1414/1000$, but neither of these fractions has a square that is exactly 2. But maybe we just haven't looked hard enough? The following theorem says that such a search would be futile. The proof, as is usually the case for theorems about irrational numbers, is by contradiction. In the proof below, we will exploit the fact that every fraction can be reduced until it is in *lowest terms*, where the numerator and denominator share no common divisors bigger than 1.

Theorem: $\sqrt{2}$ is irrational.

Proof: Suppose, to the contrary, that $\sqrt{2}$ is rational. Then there must exist positive integers a and b for which

$$\sqrt{2} = a/b$$

where a/b is a fraction in lowest terms. If we square both sides of this equation, we have

$$2 = a^2/b^2$$

or equivalently,

$$a^2 = 2b^2$$

But this implies that a^2 must be an *even* integer. And if a^2 is even, then a must also be even (since we previously showed that if a were odd, then a times itself would be odd). Thus $a = 2k$, for some integer k. Substituting that information into the equation above tells us that

$$(2k)^2 = 2b^2$$

So
$$4k^2 = 2b^2$$

which means that
$$b^2 = 2k^2$$

and therefore b^2 is an even number. And since b^2 is even, then b must be even too. But wait a second! We've just shown that a and b are both even, and this contradicts the assumption that a/b is in lowest terms. Hence the assumption that $\sqrt{2}$ is rational leads to trouble, so we are forced to conclude that $\sqrt{2}$ is irrational.   ☺

I really love this proof (as evidenced by the smiley) since it proves a very surprising result through the power of pure logic. As we will see in Chapter 12, irrational numbers are hardly rare. In fact, in a very real sense, virtually all real numbers are irrational, even though we mostly work with rational numbers in our daily lives.

Here is a fun corollary to the previous theorem. (A *corollary* is a theorem that comes as a consequence of an earlier theorem.) It takes advantage of the **law of exponentiation**, which says that for any positive numbers a, b, c,
$$\left(a^b\right)^c = a^{bc}$$

For example, this says that $\left(5^3\right)^2 = 5^6$, which makes sense, since
$$\left(5^3\right)^2 = (5 \times 5 \times 5) \times (5 \times 5 \times 5) = 5^6$$

Corollary: There exist irrational numbers a and b such that a^b is rational.

It's pretty cool that we can prove this theorem right now, even though we currently know only one irrational number, namely $\sqrt{2}$. We call the following proof an *existence proof*, since it will show you that such values of a and b exist, without telling you what a and b actually are.

Proof: We know that $\sqrt{2}$ is irrational, so consider the number $\sqrt{2}^{\sqrt{2}}$. Is this number rational? If the answer is yes, then the proof is done, by letting $a = \sqrt{2}$ and $b = \sqrt{2}$. If the answer is no, then we now have a new irrational number to play with, namely $\sqrt{2}^{\sqrt{2}}$. So if we let $a = \sqrt{2}^{\sqrt{2}}$ and $b = \sqrt{2}$, then, using the law of exponentiation, we get

$$a^b = \left(\sqrt{2}^{\sqrt{2}}\right)^{\sqrt{2}} = \sqrt{2}^{\sqrt{2} \times \sqrt{2}} = \sqrt{2}^2 = 2$$

which is rational. Thus, regardless of whether $\sqrt{2}^{\sqrt{2}}$ is rational or irrational, we can find a and b such that a^b is rational.   ☺

Existence proofs like the one above are often clever, but sometimes a little unsatisfying, since they might not tell you all the information that you are seeking. (By the way, if you are curious, the number $\sqrt{2}^{\sqrt{2}}$ is irrational, but that's beyond the scope of this chapter.)

More satisfying are *constructive proofs*, which show you exactly how to find the information you want. For example, one can show that every rational number a/b must either terminate or repeat (since in the long division process, eventually b must divide a number that it has previously divided). But is the reverse statement true? Certainly a terminating decimal must be a rational number. For example, $0.12358 = 12{,}358/100{,}000$. But what about repeating decimals? For example, must the number $0.123123123\ldots$ necessarily be a rational number? The answer is yes, and here is a clever way to find exactly what rational number it is. Let's give our mystery number a name, like w (as in waltz), so that

$$w = 0.123123123\ldots$$

Multiplying both sides by 1000 gives us

$$1000w = 123.123123123\ldots$$

Subtracting the first equation from the second, we get

$$999w = 123$$

and therefore

$$w = \frac{123}{999} = \frac{41}{333}$$

Let's try this with another repeating decimal, where we don't start repeating from the very first digit. What fraction is represented by the decimal expansion $0.83333\ldots$? Here we let

$$x = 0.83333\ldots$$

Therefore

$$100x = 83.3333\ldots$$

and

$$10x = 8.3333\ldots$$

When we subtract $10x$ from $100x$ everything after the decimal point cancels, leaving us with

$$90x = (83.3333\ldots) - (8.3333\ldots) = 75$$

Therefore

$$x = \frac{75}{90} = \frac{5}{6}$$

Using this procedure, we can constructively prove that a number is rational if and only if its decimal expansion is either terminating or repeating. If the number has an infinite decimal expansion that doesn't repeat, like

$$v = .1234567891011121314 15\ldots$$

then that number is irrational.

Proofs by Induction

Let's go back to proving theorems about positive integers. In Chapter 1, after observing that

$$
\begin{aligned}
1 &= 1 \\
1 + 3 &= 4 \\
1 + 3 + 5 &= 9 \\
1 + 3 + 5 + 7 &= 16
\end{aligned}
$$

we suspected, and later proved, that the sum of the first n odd numbers is n^2. We accomplished this by a clever *combinatorial proof* where we counted the squares of a checkerboard in two different ways. But let's try a different approach that doesn't require as much cleverness. Suppose I *told* you, or you accepted on faith, that the sum of the first 10 odd numbers $1 + 3 + \cdots + 19$ has a total of $10^2 = 100$. If you accepted that statement, then it would automatically follow that when we added the 11th odd number, 21, the total would be 121, which is 11^2. In other words, the truth of the statement for 10 terms automatically implies the truth of the statement for 11 terms. That's the idea of a *proof by induction*. We show that some statement involving n is true at the beginning (usually the statement when $n = 1$), then we show that *if* the theorem is true when $n = k$, then it will automatically continue to be true when $n = k + 1$. This forces the statement to be true for all values of n. Proofs by induction are analogous to climbing a ladder: suppose we show that you can step onto the ladder and that, if you have already made it to one step, you can always make it to the next. A bit of thought should convince you that you could get to any step on the ladder.

For example, with the sum of the first n odd numbers, our goal is to show that for all $n \geq 1$,

$$1 + 3 + 5 + \cdots + (2n - 1) = n^2$$

We see that the sum of just the first odd number, 1, is indeed 1^2, so the statement is certainly true when $n = 1$. Next we notice that *if* the sum of the first k odd numbers is k^2, namely,

$$1 + 3 + 5 + \cdots + (2k - 1) = k^2$$

then when we add the next odd number (namely $2k + 1$), we see that

$$
\begin{aligned}
1 + 3 + 5 + \cdots + (2k - 1) + (2k + 1) &= k^2 + (2k + 1) \\
&= (k + 1)^2
\end{aligned}
$$

In other words, if the sum of the first k odd numbers is k^2, then the sum of the first $k + 1$ odd numbers is guaranteed to equal $(k + 1)^2$. Thus, since the theorem is true for $n = 1$, it will continue to be true for all values of n. □

Induction is a powerful tool. The very first problem we looked at in this book was to consider the sum of the first n numbers. We showed, by various means, that

$$1 + 2 + 3 + \cdots + n = \frac{n(n + 1)}{2}$$

That statement is certainly true when $n = 1$ (since $1 = 1(2)/2$). And if we assume that the statement is true for some number k:

$$1 + 2 + 3 + \cdots + k = \frac{k(k + 1)}{2}$$

then when we add the $(k + 1)$st number to that sum, we get

$$
\begin{aligned}
1 + 2 + 3 + \cdots + k + (k + 1) &= \frac{k(k + 1)}{2} + (k + 1) \\
&= (k + 1)\left(\frac{k}{2} + 1\right) \\
&= \frac{(k + 1)(k + 2)}{2}
\end{aligned}
$$

This is our formula with n replaced by $k + 1$. Thus if the formula is valid when $n = k$ (where k can be *any* positive number), then it will continue to be valid when $n = k + 1$. Hence the identity holds for all positive values of n. □

We will see more examples of induction in this chapter and later in this book. But to help reinforce it further, here's a song written by "mathemusicians" Dane Camp and Larry Lesser. It's sung to the tune of "Blowin' in the Wind" by Bob Dylan.

> How can you tell that a statement is true
> For every value of n?
> Well there's just no way you can try them all.
> Why you could just barely begin!
> Is there a tool that can help us resolve
> This infinite quand'ry we're in?
>
> The answer, my friend, is knowin' induction.
> The answer is knowin' induction!
>
> First you must find an initial case
> For which the statement is true,
> Then you must show if it's true for k
> Then $k + 1$ must work too!
> Then all statements fall just like dominos do.
> Tell me how did we score such a coup?
>
> The answer, my friend, is knowin' induction.
> The answer is knowin' induction!
>
> If I told you n times, I told you $n + 1$,
> The answer is knowin' induction!

⧓ **Aside**

In Chapter 5, we discovered several relationships with Fibonacci numbers. Let's see how to prove some of those identities using induction.

Theorem: For $n \geq 1$,

$$F_1 + F_2 + \cdots + F_n = F_{n+2} - 1$$

Proof (by induction): When $n = 1$, this says $F_1 = F_3 - 1$, which is the same as $1 = 2 - 1$, which is certainly true. Now assume that theorem is true when $n = k$. That is,

$$F_1 + F_2 + \cdots + F_k = F_{k+2} - 1$$

Now when we add the next Fibonacci number F_{k+1} to both sides, we get

$$\begin{aligned}
F_1 + F_2 + \cdots + F_k + F_{k+1} &= F_{k+1} + F_{k+2} - 1 \\
&= F_{k+3} - 1
\end{aligned}$$

as desired. □

The proof for the sum of the squares of Fibonacci numbers is equally simple.

Theorem: For $n \geq 1$,

$$F_1^2 + F_2^2 + \cdots + F_n^2 = F_n F_{n+1}$$

Proof (by induction): When $n = 1$, this says $F_1^2 = F_1 F_2$, which is true, since $F_2 = F_1 = 1$. Assuming the theorem when $n = k$, we have

$$F_1^2 + F_2^2 + \cdots + F_k^2 = F_k F_{k+1}$$

Adding F_{k+1}^2 to both sides gives us

$$\begin{aligned}
F_1^2 + F_2^2 + \cdots + F_k^2 + F_{k+1}^2 &= F_k F_{k+1} + F_{k+1}^2 \\
&= F_{k+1}\left(F_k + F_{k+1}\right) \\
&= F_{k+1} F_{k+2}
\end{aligned}$$

as desired. □

In Chapter 1, we noticed that "the sum of the cubes is the square of the sum." That is,

$$1^3 = 1^2$$
$$1^3 + 2^3 = (1 + 2)^2$$
$$1^3 + 2^3 + 3^3 = (1 + 2 + 3)^2$$
$$1^3 + 2^3 + 3^3 + 4^3 = (1 + 2 + 3 + 4)^2$$

but we weren't ready to prove it. We can now do it pretty quickly by induction. The general pattern says that for $n \geq 1$,

$$1^3 + 2^3 + 3^3 + \cdots + n^3 = (1 + 2 + 3 + \cdots + n)^2$$

and since we already know that $1 + 2 + \cdots + n = \frac{n(n+1)}{2}$, we prove the equivalent theorem, shown below.

Theorem: For $n \geq 1$,

$$1^3 + 2^3 + 3^3 + \cdots + n^3 = \frac{n^2(n+1)^2}{4}$$

Proof (by induction): When $n = 1$, the theorem gives us the true statement $1^3 = 1^2(2^2)/4$. Inductively, if the theorem is true when $n = k$ so that

$$1^3 + 2^3 + 3^3 + \cdots + k^3 = \frac{k^2(k+1)^2}{4}$$

then when we add $(k+1)^3$ to both sides, we get

$$
\begin{aligned}
1^3 + 2^3 + 3^3 + \cdots + k^3 + (k+1)^3 &= \frac{k^2(k+1)^2}{4} + (k+1)^3 \\
&= (k+1)^2 \left(\frac{k^2}{4} + (k+1) \right) \\
&= (k+1)^2 \left(\frac{k^2 + 4(k+1)}{4} \right) \\
&= \frac{(k+1)^2(k+2)^2}{4}
\end{aligned}
$$

as desired.   □

✂ Aside

Here is a geometrical proof of the sum of the cubes identity.

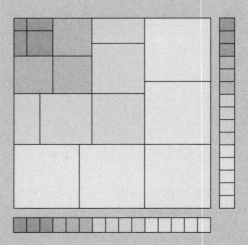

Let's compute the area of this figure in two different ways, then compare answers. On the one hand, since the figure is a square where each side has length $1 + 2 + 3 + 4 + 5$, its area is $(1 + 2 + 3 + 4 + 5)^2$.

On the other hand, if we start in the upper left corner and move diagonally downward, we can see one 1-by-1 square, then two 2-by-2 squares (with one square cut into two halves), then three 3-by-3 squares, then four 4-by-4 squares (with one square cut into two halves), then five 5-by-5 squares. Consequently, the area of this figure is equal to

$$(1 \times 1^2) + (2 \times 2^2) + (3 \times 3^2) + (4 \times 4^2) + (5 \times 5^2)$$
$$= 1^3 + 2^3 + 3^3 + 4^3 + 5^3$$

Since both areas must be equal, it follows that

$$1^3 + 2^3 + 3^3 + 4^3 + 5^3 = (1 + 2 + 3 + 4 + 5)^2$$

The same sort of picture can be created for the square with side lengths $1 + 2 + \cdots + n$ to establish

$$1^3 + 2^3 + 3^3 + \cdots + n^3 = (1 + 2 + 3 + \cdots + n)^2$$

☺

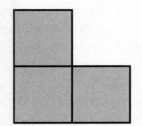

Proofs by induction can handle more than just sums. Induction is often useful whenever the solution to your "larger" problem (of size $k + 1$) can be expressed in terms of a smaller problem (of size k). Here is my favorite proof by induction, and it relates to the checkerboard tiling problem at the beginning of the chapter. But instead of proving that something is impossible, we use it to show that something is always possible. And instead of tiling the board with dominos, we tile the board with *trominos*, which are little L-shaped tiles that cover 3 squares.

Since 64 is not a multiple of 3, we can't cover an 8-by-8 checkerboard with trominos alone. But if you place a 1-by-1 square on the board, then we claim that the rest of the board can be tiled with trominos regardless of the square's location. In fact, not only is this statement true for 8-by-8 checkerboards, it's also true for boards of size 2-by-2, 4-by-4, 16-by-16, and so on.

Theorem: For all $n \geq 1$, a checkerboard of size 2^n-by-2^n can be tiled with non-overlapping trominos and a single 1-by-1 square, where the square can be located anywhere on the board.

Proof (by induction): The theorem is true when $n = 1$, since any 2-by-2 board can be covered with a single tromino and square, where the square can be anywhere on the board. Now suppose the theorem holds when $n = k$, that is, for boards of size 2^k-by-2^k. Our goal is to show that it remains true for boards of size 2^{k+1}-by-2^{k+1}. Place the 1-by-1 square anywhere on the board, then divide the board into four quadrants, as illustrated below.

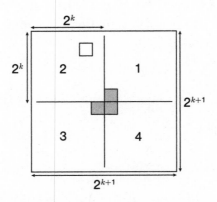

Covering a checkerboard with trominos

Now the quadrant containing the square has size 2^k-by-2^k, so it can be tiled with trominos (by the assumption that the theorem is true when $n = k$). Next we place a tromino in the center of the board, which intersects the *other* three quadrants. These quadrants have size 2^k-by-2^k with one square specified, so they too can be covered with non-overlapping trominos. This provides us with the desired tiling of the original 2^{k+1}-by-2^{k+1} board.   ☺

The final identity we prove in this section has many useful applications. We'll prove it by induction, as well as a few other ways. The motivating question is: What do you get when you add up the first n powers of 2, beginning with $2^0 = 1$? Here are the first few powers of 2:

$$1, 2, 4, 8, 16, 32, 64, 128, 256, 512, 1024, \ldots$$

As we start to add these numbers together, we see:

$$
\begin{aligned}
1 &= 1 \\
1+2 &= 3 \\
1+2+4 &= 7 \\
1+2+4+8 &= 15 \\
1+2+4+8+16 &= 31
\end{aligned}
$$

Do you see the pattern? Each sum is 1 less than the next higher power of 2. The general formula is as follows:

Theorem: For $n \geq 1$,

$$1 + 2 + 4 + 8 + \cdots + 2^{n-1} = 2^n - 1$$

Proof (by induction): As noted above, the theorem is true when $n = 1$ (and 2, 3, 4, and 5, for that matter!). Assuming the theorem holds for $n = k$, we can say that

$$1 + 2 + 4 + 8 + \cdots + 2^{k-1} = 2^k - 1$$

When we add the next power of 2, which is 2^k, to both sides, we get

$$
\begin{aligned}
1 + 2 + 4 + 8 + \cdots + 2^{k-1} + 2^k &= (2^k - 1) + 2^k \\
&= 2 \cdot 2^k - 1 \\
&= 2^{k+1} - 1 \qquad \square
\end{aligned}
$$

In Chapters 4 and 5, we proved many relationships by answering a counting question in two different ways. You might say that the following combinatorial proof counts the most!

Question: How many ways can a hockey team with n players (with jerseys numbered 1 through n) choose a delegation to attend a conference where at least one player must be on the delegation?

Answer 1: Each player has 2 choices, either attend or not, so the answer would seem to be 2^n, but we need to subtract 1 to exclude the possibility that all players choose not to attend. Thus there are $2^n - 1$ possibilities.

Answer 2: Consider the largest jersey number who attends the conference. There is only 1 delegation where 1 is the largest jersey number. There are 2 delegations where 2 is the largest jersey number (since player 2 either attends alone or with player 1). There are 4 delegations where 3 is the largest jersey number (since player 3 must attend and players 1 and 2 each have two choices). Continuing in this manner, there are 2^{n-1} delegations where player n is the largest jersey number, since that player must attend, but players 1 through $n-1$ each have 2 choices (to attend or not). Altogether, there are $1 + 2 + 4 + \cdots + 2^{n-1}$ possibilites.

Since answer 1 and answer 2 are both correct, then they must be equal. Therefore $1 + 2 + 4 + \cdots + 2^{n-1} = 2^n - 1$. ☺

But perhaps the simplest proof relies only on algebra. It is reminiscent of the method we used for expressing a repeated decimal as a fraction.

Proof by Algebra:

$$\text{Let } S = \quad 1 + 2 + 4 + 8 + \cdots + 2^{n-1}$$

Multiplying both sides by 2 gives us

$$2S = \quad 2 + 4 + 8 + \cdots + 2^{n-1} + 2^n$$

Subtracting the first equation from the second, we have massive cancellation except for the first term of S and the last term of $2S$. Therefore,

$$S = 2S - S = 2^n - 1 \qquad \square$$

The theorem we just proved is actually the key to the important *binary* representation of numbers, which is how computers store and manipulate numbers. The idea behind binary is that every number can be represented as a unique sum of distinct powers of 2. For example,

$$83 = 64 + 16 + 2 + 1$$

We express this in binary by replacing each power of 2 with the number 1 and each *missing* power of 2 with 0. In our example, $83 = (1 \cdot 64) + (0 \cdot 32) + (1 \cdot 16) + (0 \cdot 8) + (0 \cdot 4) + (1 \cdot 2) + (1 \cdot 1)$. Thus 83 has binary representation

$$83 = (1010011)_2$$

How do we know that every positive number can be represented this way? Let's suppose that we were able to represent all numbers from 1 to 99 with powers of 2 in a unique way. How do we know that we can represent 100 in a unique way? Let's start by taking the largest power of 2 below 100, which would be 64. (Must we include 64? Yes, because even if we chose 1, 2, 4, 8, 16, and 32, their sum would be 63, which falls short of 100.) Once we use 64, we need to use powers of 2 to reach a total of 36. And since, by assumption, we can represent 36 in a unique way using powers of 2, this gives us our unique representation of 100. (How do we represent 36? By repeating this logic, we take the largest power of 2 that we can, and continue in this manner. Thus $36 = 32 + 4$, and so $100 = 64 + 32 + 4$ has binary representation $(1100100)_2$.) We can generalize this argument (using what's called a *strong induction proof*) to show that every positive number has a unique binary representation.

Prime Numbers

In the last section, we established that every positive integer can be uniquely expressed as a sum of distinct powers of 2. In a sense, you could say that the powers of 2 are the building blocks of the positive integers under the operation of addition. In this section, we'll see that the prime numbers play a similar role with regard to multiplication: every positive integer can be uniquely expressed as a *product* of primes. Yet unlike the powers of 2, which can be easily identified and hold few mathematical surprises, the prime numbers are much trickier, and there are still many things we don't know about them.

A *prime number* is a positive integer with exactly *two* positive divisors, namely 1 and itself. Here are the first few primes.

$$2, 3, 5, 7, 11, 13, 17, 19, 23, 29, 31, 37, 41, 43, 47, 53 \ldots$$

The number 1 is not considered a prime number because it has only *one* divisor, namely 1. (There is a more significant reason why 1 is not considered prime, which we will mention shortly.) Notice that 2 is the only even prime. Some might say that makes it the *oddest* of all prime numbers!

A positive integer with three or more divisors is called *composite* since it can be *composed* into smaller factors. The first few composites are

$$4, 6, 8, 9, 10, 12, 14, 15, 16, 18, 20, 21, 22, 24, 25, 26, 27, 28, 30 \ldots$$

For example, the number 4 has exactly three divisors: 1, 2, and 4. The number 6 has four divisors: 1, 2, 3, and 6. Note the number 1 is not composite either. Mathematicians call the number 1 a *unit*, and it has the property that it is a divisor of every integer.

Every composite number can be expressed as the product of primes. Let's factor 120 into primes. We might start by writing $120 = 6 \times 20$. Now, 6 and 20 are composite, but they can be factored into primes, namely $6 = 2 \times 3$ and $20 = 2 \times 2 \times 5$. Thus,

$$120 = 2 \times 2 \times 2 \times 3 \times 5 = 2^3 \, 3^1 \, 5^1$$

Interestingly, no matter how we initially factor our number, we still wind up with the same prime factorization. This is a consequence of the *unique factorization theorem*, also known as the **fundamental theorem of arithmetic**, which states that every positive integer greater than 1 has a unique prime factorization.

By the way, the *real* reason the number 1 is not considered to be prime is that if it were, then this theorem would not be true. For example, the number 12 could be factored as $2 \times 2 \times 3$, but it could also be factored as $1 \times 1 \times 2 \times 2 \times 3$, so the factorization into primes would no longer be unique.

Once you know how a number factors, you know an awful lot about that number. When I was a kid, my favorite number was 9, but as I grew older, my favorite numbers became larger, then gradually more complex (for example, $\pi = 3.14159\ldots$, $\phi = 1.618\ldots$, $e = 2.71828\ldots$, and i, which has no decimal representation, but we will discuss that in Chapter 10). For a while, before I started experimenting with irrational numbers, my favorite number was 2520, since it was the smallest number that was divisible by all the numbers from 1 through 10. It has prime factorization

$$2520 = 2^3 \, 3^2 \, 5^1 \, 7^1$$

Once you know a number's prime factorization, you can instantly determine how many positive divisors it has. For example, any divisor of 2520 must be of the form $2^a 3^b 5^c 7^d$ where a is 0, 1, 2, or 3 (4 choices), b is 0, 1, or 2 (3 choices), c is 0 or 1 (2 choices), and d is 0 or 1 (2 choices). Thus by the rule of product, 2520 has $4 \times 3 \times 2 \times 2 = 48$ positive divisors.

✂ Aside

The proof of the fundamental theorem of arithmetic exploits the following fact about prime numbers (proved in the first chapter of any number theory textbook). If p is a prime number and p divides a product of two or more numbers, then p must be a divisor of at least one of the terms in the product. For example,

$$999{,}999 = 333 \times 3003$$

is a multiple of 11, so 11 must divide 333 or 3003. (In fact, $3003 = 11 \times 273$.) This property is not always true with composite numbers. For example, $60 = 6 \times 10$ is a multiple of 4, even though 4 does not divide 6 or 10.

To prove unique factorization, suppose the contrary, that some number had more than one prime factorization. Suppose that N was the *smallest* number that had two different prime factorizations. Say

$$p_1 p_2 \cdots p_r = N = q_1 q_2 \cdots q_s$$

where all of the p_i and q_j terms are prime. Since N is certainly a multiple of the prime p_1, then p_1 must be a divisor of one of the q_j terms. Let's say, for ease of notation, that p_1 divides q_1. Thus, since q_1 is prime, we must have $q_1 = p_1$. So if we divide the entire equation above by p_1, we get

$$p_2 \cdots p_r = \frac{N}{p_1} = q_2 \cdots q_s$$

But now the number $\frac{N}{p_1}$ has two different prime factorizations, which contradicts our assumption that N was the smallest such number. □

✗ Aside

Incidentally, there are number systems where not everything factors in a unique way. For example, on Mars, where all Martians have two heads, they only use even numbers

$$2, 4, 6, 8, 10, 12, 14, 16, 18, 20, 22, 24, 26, 28, 30, \ldots$$

In this Martian number system, a number like 6 or 10 is considered prime because it cannot be factored into smaller even numbers. In this system, the primes and composite numbers simply alternate. Every multiple of 4 is composite (since $4k = 2 \times 2k$) and all the other even numbers (like 6, 10, 14, 18, and so on), are prime, since they can't be factored into two smaller even numbers. But now consider the number 180:

$$6 \times 30 = 180 = 10 \times 18$$

Under the Martian number system, the number 180 can be factored into primes in two different ways, so prime factorization is not unique in the number system used on this planet.

Among the numbers from 1 to 100, there are exactly 25 primes. Among the next hundred numbers, there are 21 primes, then 16 primes among the hundred numbers after that. As we look at larger and larger numbers, primes tend to become rarer (but not in a predictable way—for example there are still 16 primes between 300 and 400 and 17 primes between 400 and 500). The number of primes between 1,000,000 and 1,000,100 is only 6. The fact that primes become more scarce makes sense because a large number has so many numbers below it that could potentially divide it.

We can prove that there are stretches of 100 numbers with no primes. There are even primeless collections of consecutive numbers of length 1000, or 1 million, or as long as you'd like. Let me try to convince you of this fact by instantly providing you with 99 consecutive composite numbers (although this isn't the first time that this happens). Consider the 99 consecutive numbers

$$100! + 2, 100! + 3, 100! + 4, \ldots, 100! + 100$$

Since $100! = 100 \times 99 \times 98 \times \cdots \times 3 \times 2 \times 1$, it must be divisible by all the numbers from 2 to 100. Now consider a number like $100! + 53$. Since 53 divides into 100!, then it must also divide into $100! + 53$. The same argument shows that for all $2 \leq k \leq 100$, $100! + k$ must be a multiple of k, so it must be composite.

> ✂ **Aside**
>
> Note that our argument doesn't say anything about the primeness of $100! + 1$, but we can determine that as well. There is a beautiful theorem called *Wilson's theorem*, which says that n is a prime number if and only if $(n - 1)! + 1$ is a multiple of n. Try it on a few small numbers to see it in action: $1! + 1 = 2$ is a multiple of 2; $2! + 1 = 3$ is a multiple of 3; $3! + 1 = 7$ is *not* a multiple of 4; $4! + 1 = 25$ is a multiple of 5; $5! + 1 = 121$ is *not* a multiple of 6; $6! + 1 = 721$ is a multiple of 7; and so on. Consequently, since 101 is prime, Wilson's theorem says that $100! + 1$ is a multiple of 101, and is therefore composite. Thus the numbers $100!$ through $100! + 100$ comprise 101 consecutive composite numbers.

With prime numbers becoming scarcer and scarcer among the very large numbers, it is natural to wonder if at some point we simply run out of primes. As Euclid told us over two thousand years ago, this will not be the case. But don't just take his word for it; enjoy the proof for yourself.

Theorem: There are infinitely many primes.

Proof: Suppose, to the contrary, that there were only finitely many primes. Hence there must be some *largest* prime number, which we shall denote by P. Now consider the number $P! + 1$. Since $P!$ is divisible by all numbers between 2 and P, none of those numbers can divide $P! + 1$. Thus $P! + 1$ must have a prime factor that is larger than P, contradicting the assumption that P was the largest prime.   □

Although we will never find a largest prime number, that doesn't stop mathematicians and computer scientists from searching for larger and larger primes. As of this writing, the largest *known* prime has 17,425,170 digits. Just to write that number down would require nearly a hundred books of this size. Yet we can describe that number on one line:

$$2^{57,885,161} - 1$$

The reason it has that simple form is that there are especially efficient methods for determining whether or not numbers of the form $2^n - 1$ or $2^n + 1$ are prime.

✕ Aside

The great mathematician Pierre de Fermat proved that if p is an odd prime number, then the number $2^{p-1} - 1$ must be a multiple of p. Check this out with the first few odd primes. For the primes $3, 5, 7, 11$, we see $2^2 - 1 = 3$ is a multiple of 3; $2^4 - 1 = 15$ is a multiple of 5; $2^6 - 1 = 63$ is a multiple of 7; and $2^{10} - 1 = 1023$ is a multiple of 11. As for composite numbers, it is clear that if n is even, then $2^{n-1} - 1$ must be odd, so it can't be a multiple of n. Checking the first few odd composites $9, 15,$ and 21, we see that $2^8 - 1 = 255$ is not a multiple of 9; $2^{14} - 1 = 16,383$ is not a multiple of 15; and $2^{20} - 1 = 1,048,575$ is not a multiple of 21 (nor even a multiple of 3). As a result of Fermat's theorem, if a large number N has the property that $2^{N-1} - 1$ is not a multiple of N, then we can be 100 percent sure that N is not prime, *even without knowing what the factors of N are!* However, the converse of Fermat's theorem is not true. There do exist some composite numbers (called *pseudoprimes*) that behave like primes. The smallest example is $341 = 11 \times 31$, which has the property that $2^{340} - 1$ is a multiple of 341. Although it's been shown that pseudoprimes are relatively rare, there are an infinite number of them, but there are tests to weed them out.

Prime numbers have many applications, particularly within computer science. Primes are at the heart of nearly every encryption algorithm, including public key cryptography, which allows for secure financial transactions across the Internet. Many of these algorithms are based on the fact that there are relatively fast ways to determine if a number is prime or not, but there are no known fast ways of factoring large numbers. For example, if I multiplied two random 1000-digit primes together and gave you their 2000-digit answer, it is highly unlikely that any human or computer (unless a quantum computer is built someday) could determine the original prime numbers. Codes that are based on our inability to factor large numbers (such as the RSA method) are believed to be quite secure.

People have been fascinated with prime numbers for thousands of years. The ancient Greeks said that a number is *perfect* if it is equal to the sum of all of its proper divisors (every divisor except itself). For example, 6 is perfect since it has proper divisors 1, 2, and 3, which sum to 6. The next perfect number is 28, which has proper divisors 1, 2, 4, 7, and 14, which sum to 28. The next two perfect numbers are 496 and 8128. Is there any pattern here? Let's look at their prime factorizations.

$$
\begin{aligned}
6 &= 2 \times 3 \\
28 &= 4 \times 7 \\
496 &= 16 \times 31 \\
8128 &= 64 \times 127
\end{aligned}
$$

Do you see the pattern? The first number is a power of 2. The second number is one less than twice that power of 2, and it's prime. (That's why you don't see 8×15 or 32×63 on the list, since 15 and 63 are not prime.) We can summarize this pattern in the following theorem.

Theorem: If $2^n - 1$ is prime, then $2^{n-1} \times (2^n - 1)$ is perfect.

✗ Aside

Proof: Let $p = 2^n - 1$ be a prime number. Our goal is to show that $2^{n-1}p$ is perfect. What are the proper divisors of $2^{n-1}p$? The divisors that do not use the factor p are simply $1, 2, 4, 8, \ldots, 2^{n-1}$, which has sum $2^n - 1 = p$. The other proper divisors (which excludes $2^{n-1}p$) utilize the factor p, so these divisors sum to $p(1 + 2 + 4 + 8 + \cdots + 2^{n-2}) = p(2^{n-1} - 1)$. Hence the grand total of proper divisors is

$$
p + p(2^{n-1} - 1) = p(1 + (2^{n-1} - 1)) = 2^{n-1}p
$$

as desired.   □

The great mathematician Leonhard Euler (1707–1783) proved that every even perfect number is of this form. As of this writing, there have been forty-eight discovered perfect numbers, all of which are even. Are there any odd perfect numbers? Presently, nobody knows the answer to that question. It has been shown that *if* an odd perfect number exists, then it would have to contain over three hundred digits, but nobody has yet proved that they are impossible.

There are many easily stated unsolved problems pertaining to prime numbers. We have already stated that it is unknown whether there are infinitely many prime Fibonacci numbers. (It has been shown that there are only two perfect squares among the Fibonacci numbers (1 and 144) and only two perfect cubes (1 and 8).) Another unsolved problem is known as *Goldbach's conjecture*, which speculates that every even number greater than 2 is the sum of two primes. Here, too, nobody has been able to prove this conjecture, but it has been proved that if a counterexample exists, then it must have at least 19 digits. (A breakthrough was recently made on a similar-sounding problem. In 2013, Harald Helfgott proved that every odd number bigger than 7 is the sum of at most three odd primes.) Finally, we define *twin primes* to be any two prime numbers that differ by 2. The first examples of twin primes are 3 and 5, 5 and 7, 11 and 13, 17 and 19, 29 and 31, and so on. Can you see why 3, 5, and 7 are the only "prime triplets"? And even though it has been proved (as a special case of a theorem due to Gustav Dirichlet) that there are infinitely many primes that end in 1 (or end in 3 or end in 7 or end in 9), the question of whether there exists an infinite number of twin primes remains open.

Let's end this chapter with a proof that's a little fishy, but I hope you agree with the statement anyway.

Claim: All positive integers are interesting!

Proof?: You'll agree that the first few positive numbers are all very interesting. For instance, 1 is the first number, 2 is the first even number, 3 is the first odd prime, 4 is the only number that spells itself F-O-U-R, and so on. Now suppose, to the contrary, that not all numbers are interesting. Then there would have to be a first number, call it N, that was not interesting. But then that would make N interesting! Hence no uninteresting numbers exist!   ☺

CHAPTER SEVEN

The Magic of Geometry

Some Geometry Surprises

Let's begin with a geometry problem that could be presented as a magic trick. On a separate sheet of paper, follow the steps below.

Step 1. Draw a four-sided figure where the sides don't cross each other. (This is called a *quadrilateral*.) Label the four corners *A*, *B*, *C*, and *D*, in clockwise order. (See some examples below.)

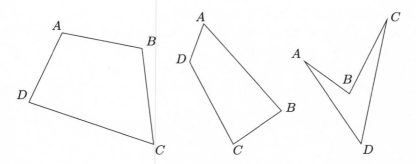

Three random quadrilaterals

Step 2. Label the midpoints of the four sides $\overline{AB}$, $\overline{BC}$, $\overline{CD}$, and $\overline{DA}$ as E, F, G, and H, respectively.

Step 3. Connect the midpoints to form a quadrilateral $EFGH$, as shown in the examples below.

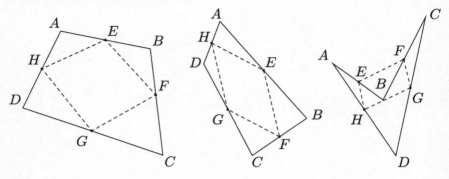

Connecting the midpoints of a quadrilateral always produces a parallelogram

Believe it or not, $EFGH$ is guaranteed to be a parallelogram. In other words, $\overline{EF}$ will be parallel to $\overline{GH}$; $\overline{FG}$ will be parallel to $\overline{HE}$. (Also, $\overline{EF}$ and $\overline{GH}$ will have the same length, as will $\overline{FG}$ and $\overline{HE}$.) This is illustrated in the figure above, but you should try some examples of your own.

Geometry is filled with surprises like this. Applying simple logical arguments to the simplest of assumptions, one often winds up with beautiful results. Let's take a short quiz to test your geometric intuition. Some of these questions have very intuitive answers, but some of these questions have answers that will surprise you, even after you have learned the appropriate geometry.

Question 1. A farmer wishes to build a rectangular fence with a perimeter of 52 feet. What should the dimensions of the rectangle be to maximize the rectangle's area?

A) A square (13 feet per side).

B) Close to the proportions of the golden ratio 1.618 (say around 16 feet by 10 feet).

C) Make the width as long as possible (close to 26 feet by 0 feet).

D) All answers above give the same area.

Question 2. Consider the parallel lines below, with X and Y on the lower line. We wish to choose a third point on the upper line so that the triangle formed by X, Y, and the upper point has the smallest perimeter. What upper point should be chosen?

A) Point A (above the point that is halfway between X and Y).

B) Point B (so the triangle formed by B, X, and Y is a right triangle).

C) As far away from X and Y as possible (like point C).

D) It doesn't matter. All triangles will have the same perimeter.

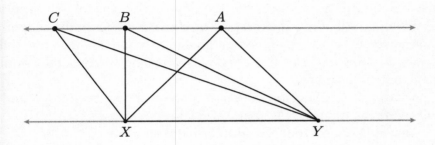

Which point on the upper line results in a triangle (with points X and Y) with the smallest perimeter? Which point results in the greatest area?

Question 3. Using the same figure as above, what point P on the upper line should be chosen so that the triangle formed by X, Y, and P has the greatest area?

A) Point A.

B) Point B.

C) As far away from X and Y as possible.

D) It doesn't matter. All triangles will have the same area.

Question 4. The distance between two goalposts on a football field is 360 feet (120 yards). A rope of length 360 feet is about to be tied tightly between the bottom of the two goalposts, when an extra foot of rope is added. How high can the rope be lifted in the middle of the field?

A) Less than one inch off the ground.

B) Just high enough to crawl under.

C) Just high enough to walk under.

D) High enough to drive a truck under.

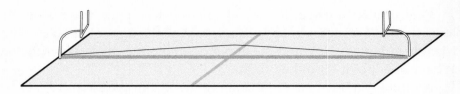

A rope of length 361 feet is tied between goalposts 360 feet apart.
How high can we lift the rope in the middle of the field?

The answers to these questions are given below. I think the first two answers are pretty intuitive, but the other two answers will surprise most people. All of the answers will be explained later in the chapter.

Answer 1. A. For any given perimeter, to maximize the area of the rectangle, you should let all sides have equal length. Thus, the optimal shape will be a square.

Answer 2. A. Choosing the point A above the midpoint of X and Y will create the triangle XAY that has smallest perimeter.

Answer 3. D. All triangles will have the same area.

Answer 4. D. At the midpoint of the field, the rope can be lifted more than 13 feet in the air—high enough for most trucks to fit under.

We can explain the first answer using simple algebra. For a rectangle with top and bottom lengths b (b as in base) and left and right lengths h (h as in height), the *perimeter* of the rectangle is $2b + 2h$, which is the sum of the lengths of all four sides. The *area* measures what can fit into the rectangle and is the product bh. (We'll say more about area later.) Since the perimeter is required to be 52 feet, we have $2b + 2h = 52$, or equivalently

$$b + h = 26$$

And since $h = 26 - b$, the area bh that we wish to maximize is equal to

$$b(26 - b) = 26b - b^2$$

What value of b maximizes this quantity? We'll see an easy way to do this with calculus in Chapter 11. But we can also find b by using the technique of completing the square, presented in Chapter 2. Notice that

$$26b - b^2 = 169 - (b^2 - 26b + 169) = 169 - (b - 13)^2$$

is the area of our rectangle when base b is chosen. When $b = 13$ our rectangle has area $169 - 0^2 = 169$. When $b \neq 13$, our area is

$$169 - (\text{something not equal to } 0)^2$$

Since we are subtracting a positive quantity from 169, this will always be less than 169. Consequently, the area of the rectangle is maximized when $b = 13$ and $h = 26 - b = 13$. One of the amazing things about geometry is that the fact that the farmer has 52 feet of fence is irrelevant. To maximize the area of a rectangle with any given perimeter p, the same technique can be used to show that the optimal shape will be a square, where all sides are of length $p/4$.

In order to explain the other problems, we need to first look at some seemingly paradoxical results and explore some classics of geometry. Why should a triangle have 180 degrees? What is the Pythagorean theorem all about? How can you tell when two triangles will have the same shape, and why should we care?

Geometry Classics

The subject of geometry goes back to the ancient Greeks. The name geometry comes from the Greek words for "earth" (*geo*) and "measurement" (*metria*), and indeed geometry's original uses were for the study of earthly measurements in surveying and construction, and for heavenly applications such as astronomy. But the ancient Greeks were masters of deductive reasoning and developed the subject into the art form that it is today. All of the results of geometry that were known at the time (around 300 BC) were compiled by Euclid into *The Elements*, which became one of the most successful textbooks of all time. This book introduced the ideas of mathematical rigor, logical deduction, the axiomatic method, and proofs, which are still utilized by mathematicians.

Euclid began with five *axioms* (also known as *postulates*), which are statements that we are supposed to accept as common sense. And once you accept these axioms, then you can, in principle, derive all geometrical truths from them. Here are Euclid's five axioms. (Actually, he stated the fifth axiom a little differently, but our axiom is equivalent to it.)

Axiom 1. Given any two points, we can connect them with a unique line segment.

Axiom 2. Line segments can be extended indefinitely in both directions to create lines.

Axiom 3. Given any points O and P, we can draw a unique circle, centered at O, where P is on the circle.

Axiom 4. All right angles measure 90 degrees.

Axiom 5. Given a line ℓ and a point P not on the line, there exists exactly one line through P that is parallel to ℓ.

✂ Aside

I should clarify that in this chapter we are working with *plane geometry* (also referred to as *Euclidean geometry*), where it is assumed that we are working on a flat surface, such as the *x-y plane*. But if we change some of the axioms, we can still get interesting and useful mathematical systems, like spherical geometry, which looks at points on a sphere. In spherical geometry, the "lines" are circles of maximum circumference (called *great circles*), and as a consequence, all lines must intersect somewhere, so parallel lines don't exist. If axiom 5 is changed so that there are always *at least two* different lines through P that are parallel to ℓ, then we get something called *hyperbolic geometry*, which has beautiful theorems all its own. Many of the brilliant prints created by the artist M. C. Escher were based on this type of geometry. The image below was created by Douglas Dunham using the rules of hyperbolic geometry, and was used with permission.

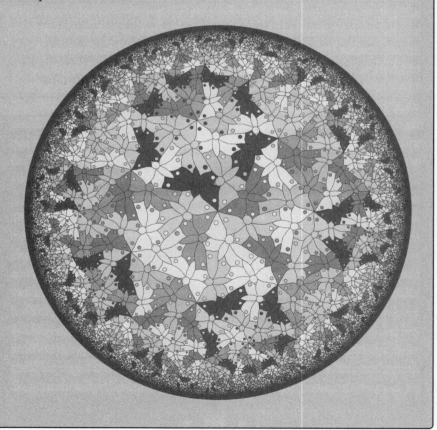

As it happens, there are other axioms that Euclid left out, and some of them will be mentioned as needed. Since this chapter is not intended to replace an entire geometry textbook, we won't try to define and prove everything from the ground up. I will assume that you have an intuitive idea about the meaning of points, lines, angles, circles, perimeters, and areas, and I will try to keep jargon and notation to a minimum, so we can concentrate on the interesting ideas of geometry.

For example, I will assume that you already know, or are willing to accept, that a circle has 360 degrees, denoted 360°. The measure of an angle is some number between 0° and 360°. Think of the hands of a clock, joined at the center of a circle. At 1 o'clock, the hands indicate 1/12th of the circle, so they form an angle of 30°. At 3 o'clock, the hands

The angles above have measures 30°, 90°, and 180°

indicate one-quarter of the circle, so they form a 90° angle. Angles of 90 degrees are called *right angles*, and the lines or segments that form them are called *perpendicular*. A straight line, such as at 6 o'clock, has an angle of 180 degrees.

Here's one piece of useful notation. The line segment connecting points A and B is denoted by $\overline{AB}$, and its length has the bar omitted, so the length of $\overline{AB}$ is AB.

When two lines cross they create four angles, as in the figure on the following page. What can we say about these angles? Notice that adjacent angles (like a and b) together form a straight line, which has 180°. Thus, angles a and b must add up to 180°. Such angles are called *supplementary*.

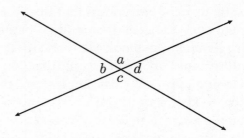

When two lines cross, the adjacent angles sum to 180°.
The non-adjacent angles (called vertical angles) are equal.
Here angles a and c form a vertical angle pair, as do angles b and d.

This property holds for every other adjacent pair of angles. That is,

$$a + b = 180°$$
$$b + c = 180°$$
$$c + d = 180°$$
$$d + a = 180°$$

Subtracting the second equation from the first equation tells us that $a - c = 0$. That is,

$$a = c$$

And subtracting the third equation from the second equation gives us

$$b = d$$

When two lines cross, the non-adjacent angles are called *vertical angles*. We have just proved the **vertical angle theorem**: vertical angles are equal.

Our next goal is to prove that the sum of the angles in any triangle is 180 degrees. To do this, we need to first say a few things about parallel lines. We say that two different lines are *parallel* if they never cross. (Remember that lines extend infinitely in both directions.) The figure opposite shows two parallel lines, ℓ_1 and ℓ_2, and a third line ℓ_3 that is not parallel to them, and therefore crosses the lines at points P and Q respectively. If you look at the picture, it appears that line ℓ_3 cuts through lines ℓ_1 and ℓ_2 at the same angle. That is, we believe that $a = e$. We call angles a and e *corresponding angles*. (Other examples of corresponding angles are angles b and f, angles c and g, and angles d and h.) It sure seems that corresponding angles should always be equal, yet this can't actually be proved as a consequence of the original five axioms. Thus, we need a new axiom.

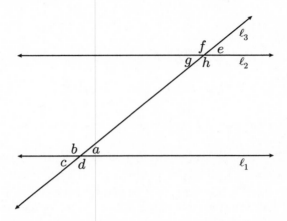

Corresponding angles are equal. Here $a = e$, $b = f$, $c = g$, and $d = h$.

Corresponding angle axiom: Corresponding angles are equal.

When combined with the Vertical Angle Theorem, this says that in the figure above, we must have

$$a = c = g = e$$

$$b = d = h = f$$

Math books give special names for some of the equal angle pairs above. For example, angles a and g, forming a Z pattern, are called *alternate interior angles*. Thus, we've shown that any angle is equal to its vertical angle, corresponding angle, and alternate interior angle. We now use this result to prove a fundamental theorem of geometry.

Theorem: For any triangle, the sum of its angles is 180 degrees.

Proof: Consider a triangle ABC like the one on the next page, with angles a, b and c. Now draw a line through the point B that is parallel to the line going through A and C.

Angles d, b, and e form a straight line, so $d + b + e = 180°$. But notice that angles a and d are alternate interior angles, as are angles c and e. Thus $d = a$ and $e = c$, so $a + b + c = 180°$, as desired.   □

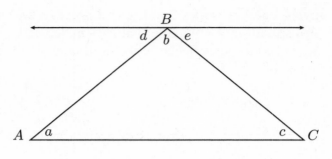

Why does $a + b + c = 180°$?

> **Aside**
>The 180° theorem for triangles is an essential fact of plane geometry, but it need not hold for other geometries. For example, consider drawing a triangle on a globe, starting at the north pole, going down to the equator along any longitudinal line, turning right at the equator, going around a quarter of the earth, then turning right again until you return to the north pole. This triangle would actually contain three right angles and sum to 270°. In spherical geometry, the sum of the angles of triangles is not constant, and the extent to which the sum of the angles exceeds 180° is directly proportional to the area of the triangle.

In geometry classes, much attention is paid to proving that various objects are *congruent*, which means that by sliding, rotating, or flipping one object, we can obtain the other object. For example, the triangles ABC and DEF pictured below are congruent, since we can slide triangle DEF so that it perfectly overlaps triangle ABC. In our figures, when two sides (or angles) have the same number of hash marks, this indicates that they have same length (or measure).

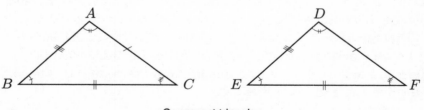

Congruent triangles

We indicate this with the $\cong$ symbol by writing $ABC \cong DEF$. This is equivalent to saying that the lengths and angles match up perfectly. Specifically, the side lengths $AB, BC,$ and CA are respectively equal to

DE, EF, and FD, and the angles associated with A, B, and C are respectively equal to the angles D, E, and F. We have marked this in the figure by putting the same symbol in angles that are equal, and similarly marked sides of the triangle that are equal.

Once you know that some of the sides and angles are equal, then the rest can follow automatically. For instance, if you know that all three sides are equal, and that two pairs of angles are equal (say $\angle A = \angle D$ and $\angle B = \angle E$), then the third pair of angles must also be equal, and so the triangles must be congruent. However, we don't even need all of this information. Once you know that two of the side lengths are equal, say $AB = DE$ and $AC = DF$, and you know that the angles *between them* are equal, here $\angle A = \angle D$, then everything else is forced: $BC = EF$, $\angle B = \angle E$, and $\angle C = \angle F$. We call this the *SAS axiom*, where SAS stands for "side-angle-side."

The SAS axiom is not a theorem, since it cannot be proved by the preceding axioms. But once we accept it, we can rigorously prove other useful theorems like SSS (side-side-side), ASA (angle-side-angle), and AAS (angle-angle-side). There is not an analogous theorem called SSA (or its embarrassing backward namesake) since the common angle must be between the two equal sides to ensure congruence. The theorem SSS is most intriguing, since it says that if two triangles have equal side lengths then they must have equal angles as well.

Let's apply SAS to prove the important isosceles triangle theorem. A triangle is *isosceles* if two of its sides have the same length. While we're at it, let's mention a few more types of triangles. An *equilateral triangle* is a triangle where all three sides have the same length. A triangle with a 90° angle is called a *right* triangle. If all angles of the triangle are less than 90°, then we have an *acute* triangle. If the triangle has an angle that is more than 90°, then it is called an *obtuse* triangle.

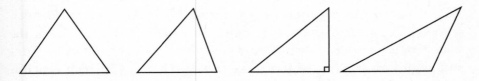

Equilateral triangle, acute triangle, right triangle, and obtuse triangle

Isosceles triangle theorem: If ABC is an isosceles triangle with equal side lengths $AB = AC$, then the angles opposite these sides must also be equal.

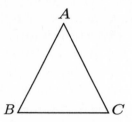

Isosceles triangle theorem: If $AB = AC$, then $\angle B = \angle C$

Proof: Begin by drawing a line from A that splits $\angle A$ into two equal angles, as in the figure below. (This is called the *angle bisector* of A.) Let's say it intersects segment $\overline{BC}$ at point X.

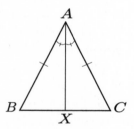

The isosceles triangle theorem can be proved by drawing an angle bisector and then applying the side-angle-side axiom to the newly created triangles

We claim that triangles BAX and CAX are congruent. This is a consequence of SAS since $BA = CA$ (by isosceles assumption), $\angle BAX = \angle CAX$ (because of the angle bisector), and $AX = AX$. (No, that's not a typo—$\overline{AX}$ appears in the two triangles we are comparing and it has the same length in both of them!) And since $BAX \cong CAX$, the *other* sides and angles must be equal too. Specifically, $\angle B = \angle C$, which was our goal.   □

✕ **Aside**

The isosceles triangle theorem can also be proved using the SSS theorem. For this proof, let M denote the *midpoint* of $\overline{BC}$, where $BM = MC$. Then draw the line segment $\overline{AM}$. Like in the previous proof, triangles BAM and CAM are congruent, since $BA = CA$ (isosceles), $AM = AM$, and $MB = MC$ (midpoint). Thus, by SSS, $BAM \cong CAM$, so their respective angles are also equal. In particular, $\angle B = \angle C$, as desired.

As a result of congruence, it follows that $\angle BAM = \angle CAM$, so the segment $\overline{AM}$ is also the angle bisector. Moreover, since $\angle BMA = \angle CMA$ and since these equal angles sum to $180°$, they must both be $90°$. So in an isosceles triangle, the angle bisector of A is also the *perpendicular bisector* of BC.

By the way, the *converse* of the isosceles triangle theorem is also true. That is, if $\angle B = \angle C$, then $AB = AC$. This can be proved by drawing the angle bisector from A to point X, as in the original proof. Now we claim that $BAX \cong CAX$ by AAS since $\angle B = \angle C$ (by assumption), $\angle BAX = \angle CAX$ (angle bisector), and $AX = AX$. Thus, $AB = AC$, so the triangle is isosceles.

In an equilateral triangle, where *all* sides are equal, we can apply the previous theorem to all three pairs of sides to see that all three angles are also equal. Thus, since the three angles must sum to $180°$, we have the following:

Corollary: In an equilateral triangle, all angles must be $60°$.

According to the SSS theorem, if two triangles ABC and DEF have matching sides ($AB = DE$, $BC = EF$, $CA = FD$), then they must also have matching angles ($\angle A = \angle D, \angle B = \angle E, \angle C = \angle F$). Is the converse statement true? If ABC and DEF have matching angles, then must they have equal sides? Certainly not, as seen in the picture below.

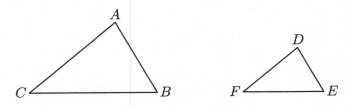

Similar triangles have matching angles and proportional sides

Two triangles with the same angle measurements are called *similar*. If two triangles ABC and DEF are similar (denoted $\triangle ABC \sim \triangle DEF$ or just $ABC \sim DEF$), then $\angle A = \angle D$, $\angle B = \angle E$, $\angle C = \angle F$. Essentially, similar triangles are just scaled versions of one another. So if $ABC \sim DEF$, then the sides must be scaled by some positive factor k. That is, $DE = kAB$, $EF = kBC$, and $FD = kCA$.

Let's apply what we have learned so far to answer question 2 from the beginning of the chapter. Recall that we started with two parallel lines with segment $\overline{XY}$ on the lower line. Our goal was to find a point P on the upper line so that triangle XYP has the smallest possible perimeter. We claimed that the following was true:

Theorem: The point P on the upper line that minimizes the perimeter of XYP is directly above the midpoint of $\overline{XY}$.

Although this problem could be solved using tricky calculus, we will see how geometry allows us to solve the problem with just a little bit of "reflection." (The proof that follows is interesting, but somewhat long, so feel free to skim or skip it.)

Proof: Let P be any point on the upper line, and let Z denote the point on the upper line that is directly above Y. (More precisely, the point Z is placed so that it is on the line containing Y, and $\overline{YZ}$ is perpendicular to both the upper and lower lines. See the figure below.) Let Y' be the point on the perpendicular line such that $Y'Z = ZY$. In other words, if the upper line were a big mirror, then Y' would be the reflection of Y through the point Z.

I claim that triangles PZY and PZY' are congruent. That's because $PZ = PZ$, $\angle PZY = 90° = \angle PZY'$, and $ZY = ZY'$, so the triangles are congruent by SAS. Consequently, $PY = PY'$, which we can exploit.

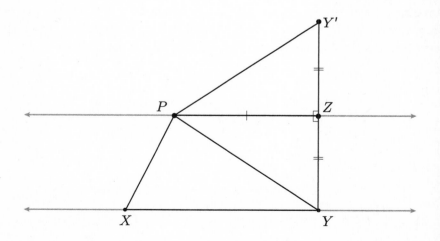

Since triangles PZY and PZY' are congruent (by SAS), we must have $PY = PY'$

The perimeter of triangle YXP is the sum of the three lengths

$$YX + XP + PY$$

and since we have shown that $PY = PY'$, the perimeter also equals

$$YX + XP + PY'$$

Now, the length YX does not depend on P, so our problem reduces to finding the point P that will minimize $XP + PY'$.

Notice that the line segments $\overline{XP}$ and $\overline{PY'}$ form a crooked path from X to Y'. Since the shortest distance between two points is a straight line, the optimal point P^* can be found by drawing a straight line from X to Y'; P^* is the point where this line intersects the upper line. See the figure below. So why aren't we done? To complete the proof we need to show that P^* is directly above the midpont of $\overline{XY}$.

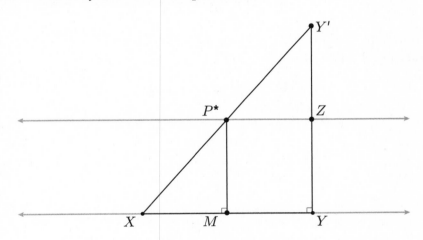

Triangles MXP^* and YXY' are similar with scale factor 2

Let M denote the point directly below P^* (so that $\overline{P^*M}$ is perpendicular to $\overline{XY}$). Since the upper and lower lines are parallel, the lengths P^*M and ZY must be equal. (This makes intuitive sense since parallel lines have a constant distance between them, but it can also be proved by drawing the segment $\overline{MZ}$ and verifying that triangles MYZ and ZP^*M are congruent by AAS.)

To prove that M is the midpoint of $\overline{XY}$, we first prove that triangles MXP^* and YXY' are similar. Notice that $\angle MXP^*$ and $\angle YXY'$ are the same angle, $\angle P^*MX = \angle Y'YX$ since they are both right angles, and once we have two matching angles, then the third angle must match as well, since the sum of the angles is 180°. What is the scale factor between these similar triangles? By construction, the length

$$YY' = YZ + ZY' = 2YZ = 2MP^*$$

so the scale factor is 2. Consequently, the length XM is half the length of XY, and is therefore the midpoint of $\overline{XY}$.

Summarizing, we have shown that the point P^* on the upper line that minimizes the perimeter of triangle XYP lies directly above the midpoint of $\overline{XY}$.   □

Sometimes geometry problems can be solved using algebra. For example, suppose the line segment $\overline{AB}$ is drawn on the plane where A has coordinates (a_1, a_2) and B has coordinates (b_1, b_2). Then the midpoint M, which is halfway between A and B, has coordinates

$$M = \left(\frac{a_1 + b_1}{2}, \frac{a_2 + b_2}{2} \right)$$

as illustrated below. For example, if $A = (1, 2)$ and $B = (3, 4)$, then the midpoint of $\overline{AB}$ is $M = ((1+3)/2, (2+4)/2) = (2, 3)$.

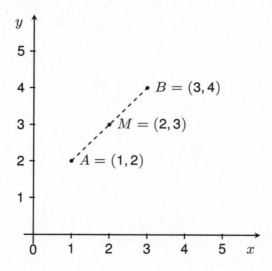

The midpoint of a line segment can be found by taking the average of its endpoints

Let's use this idea to prove a useful fact about triangles. Draw a triangle, then connect the midpoints of any two sides with a line segment. Do you notice anything interesting? The answer is given in the following theorem.

Triangle midpoint theorem: Given any triangle ABC, if we draw a line segment connecting the midpoint of $\overline{AB}$ with the midpoint of $\overline{BC}$, then that line segment will be parallel to the third side $\overline{AC}$. Moreover, if the length of $\overline{AC}$ is b, then the segment connecting the midpoints will have length $b/2$.

Proof: Place the triangle ABC on the plane in such a way that point A is at the origin $(0,0)$ and side AC is horizontal so that point C is at the point $(b,0)$, as shown below. Suppose that point B is located at (x,y). Then the midpoint of $\overline{AB}$ has coordinates $(x/2, y/2)$ and the midpoint of $\overline{BC}$ will have coordinates $((x+b)/2, y/2)$. Since both midpoints have the same y-coordinate, the line segment connecting them must be horizontal, so it will be parallel to side $\overline{AC}$. Moreover the length of this line segment is $(x+b)/2 - x/2 = b/2$, as desired. □

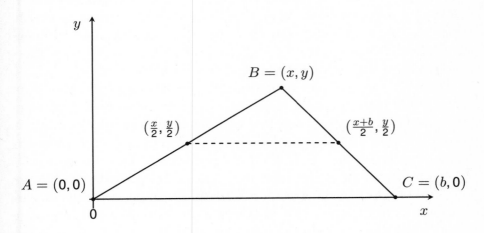

When the midpoints of two sides of a triangle are connected by a line segment, then that segment is parallel to the third side and has half of its length

The triangle midpoint theorem reveals the secret to the magic trick at the beginning of this chapter. Starting with quadrilateral $ABCD$, we connected the midpoints to form a second quadrilateral $EFGH$, which always turned out to be a parallelogram. Let's see why this works. If we imagine a diagonal line drawn from vertex A to vertex C, then this creates two triangles ABC and ADC, as shown on the next page.

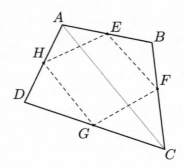

By the triangle midpoint theorem, $\overline{EF}$ and $\overline{GH}$ are both parallel to $\overline{AC}$

Applying the triangle midpoint theorem to triangles ABC and ADC, we find that $\overline{EF}$ will be parallel to $\overline{AC}$, and $\overline{AC}$ will be parallel to $\overline{GH}$. Thus $\overline{EF}$ is parallel to $\overline{GH}$. (Moreover, $\overline{EF}$ and $\overline{GH}$ have the same length, since they are both half the length of $\overline{AC}$.) By the same logic, by imagining the diagonal line from B to D, we get that $\overline{FG}$ and $\overline{HE}$ are parallel and have the same length. Thus EFGH is a parallelogram.

Many of the previous theorems dealt with triangles, and indeed in geometry a great deal of time is spent studying triangles. Triangles are the simplest of *polygons*, followed by *quadrilaterals* (4-sided polygons), *pentagons* (5-sided polygons), and so on. A polygon with n sides is sometimes called an *n-gon*. We have proved that the sum of the angles of any triangle is 180°. What can be said of polygons with more than three sides? A quadrilateral, such as a square, rectangle, or parallelogram, has 4 sides. In a rectangle, all four angles have a measure of 90°, so the sum of the angles must be 360°. The next theorem shows that is true for *any* quadrilateral. You might call this *a 4-gon conclusion*. (Sorry, I couldn't resist!)

Theorem: The sum of the angles of a quadrilateral is 360°.

Proof: Take any quadrilateral with vertices A, B, C, D like the one on the next page. By drawing a line segment from A to C, the quadrilateral is broken into two triangles, each of which has an angle sum of 180°. Hence the angles of the quadrilateral sum to $2 \times 180° = 360°$.   □

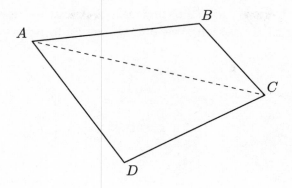

The sum of the angles of a quadrilateral is 360°

One more theorem should reveal the general pattern.

Theorem: The sum of the angles of a pentagon is 540°.

Proof: Consider any pentagon with vertices A, B, C, D, E like the one below. By drawing a line segment from A to C, the pentagon is broken into a triangle and a quadrilateral. We know that the angles of triangle ABC sum to 180° and by our 4-gon conclusion, the sum of the angles of quadrilateral $ACDE$ is 360°. Hence the sum of the angles of a pentagon is $180° + 360° = 540°$. ☐

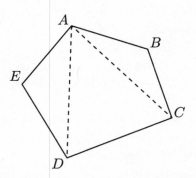

The sum of the angles of a pentagon is 540°

We obtain the following theorem by repeating this argument for an n-gon by doing a proof by induction or by creating $n - 2$ triangles by drawing line segments from A to all other vertices.

Theorem: The sum of the angles of an n-gon is $180(n - 2)$ degrees.

Here's a magical application of this theorem. Draw an *octagon* (an 8-sided polygon) and put 5 points anywhere inside it. Then connect the vertices and points in such a way that the only shapes inside the octagon are triangles. (This is called a *triangulation*.) Below are two examples of different triangulations, with one empty for you to try yourself.

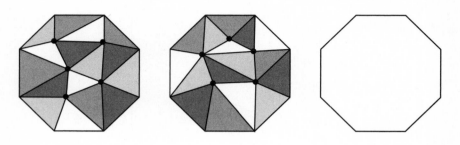

In both of my examples, we ended up with exactly 16 triangles. In the third octagon, no matter where you place the 5 points inside it, you should also have exactly 16 triangles, if everything was done properly. (If you didn't get 16 triangles, look closely at every interior region and make sure that it has only three points. If a triangular-looking region has four points, then you need to insert another line segment to properly split it into two triangles.) The explanation for this comes from the following theorem.

Theorem: Any triangulation of a polygon with n sides and p interior points will contain exactly $2p + n - 2$ triangles.

In the previous example, $n = 8$ and $p = 5$, so the theorem predicts exactly $10 + 8 - 2 = 16$ triangles.

Proof: Suppose the triangulation has exactly T triangles. We prove that $T = 2p + n - 2$ by answering the following counting question in two ways.

Question: What is the sum of the angles of all of the triangles?

Answer 1: Since there are T triangles, each with an angle sum of $180°$, the sum must be $180T$ degrees.

Answer 2: Let's break the answer into two cases. The angles surrounding each of the p interior points must go all around the circle, so they contribute $360p$ to the total. On the other hand, from our previous theorem, we know that the sum of the angles *on* the n-gon itself is $180(n - 2)$ degrees. Hence the sum of all the angles is $360p + 180(n - 2)$ degrees.

Equating our two answers gives us

$$180T = 360p + 180(n - 2)$$

Dividing both sides by 180 gives us

$$T = 2p + n - 2$$

as predicted.   ☺

Perimeters and Areas

The *perimeter* of a polygon is the sum of the lengths of its sides. For example, in a rectangle with a base of length b and a height of length h, its perimeter would be $2b + 2h$, since it has two sides of length b and two sides of length h. What about the area of the rectangle? We define the *area* of a 1-by-1 square (the *unit square*) to have area 1. When b and h are positive integers, like in the figure below, we can break up the region into bh 1-by-1 squares, so its area is bh. In general, for any rectangle with base b and height h (where b and h are positive, but not necessarily integers), we define its area to be bh.

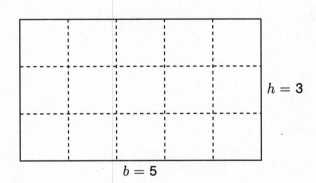

A rectangle with base b and height h has perimeter $2b + 2h$ and area bh

✂ Aside

Throughout this chapter, we have used algebra to help us explain geometry. But sometimes geometry can help us explain algebra too. Consider the following algebra problem. How small can the quantity $x + \frac{1}{x}$ be, where x is allowed to be any positive number? When $x = 1$, we get 2; when $x = 1.25$, we get $1.25 + 0.8 = 2.05$; when $x = 2$, we get 2.5. The data seem to suggest that the smallest answer we can get is 2, and that's true, but how can we be sure? We'll find a straightforward way to answer this question with calculus in Chapter 11, but with a little cleverness we can solve this problem with simple geometry.

Consider the geometric object below made up of four dominos, each with dimensions x by $1/x$, strung together to form a square with a hole in the middle of it. What is the area of the whole region, including the hole?

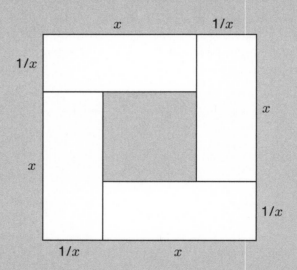

On the one hand, since the region is a square with side length $x + 1/x$, its area must be $(x + 1/x)^2$. On the other hand, the area of each domino is 1, so the area of the region must be at least 4. Consequently

$$(x + 1/x)^2 \geq 4$$

which implies that $x + 1/x \geq 2$, as desired.   ☺

Starting with the area of a rectangle, it is possible to derive the area of just about any geometrical figure. First and foremost, we derive the area of a triangle.

Theorem: A triangle with base length b and height h has area $\frac{1}{2}bh$.

To illustrate, all three triangles shown below have the same base length b and the same height h, and therefore they all have the same area. This was essentially the content of question 3 at the beginning of the chapter, and it comes as a surprise to many.

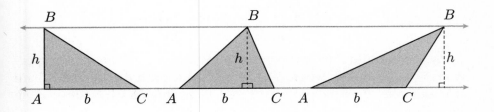

The area of a triangle with base b and height h is $\frac{1}{2}bh$.
This is true, regardless of whether the triangle is right-angled, acute, or obtuse.

There are three cases to consider, depending on the size of the base angles $\angle A$ and $\angle C$. If $\angle A$ or $\angle C$ is a right angle, then we can make a copy of triangle ABC and put the two triangles together to form a rectangle of area bh, as is shown below. Since triangle ABC occupies half of the rectangle's area, then the triangle must have area $\frac{1}{2}bh$, as claimed.

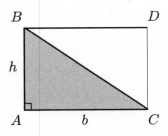

Two right triangles of base b and height h can form a rectangle of area bh

When $\angle A$ and $\angle C$ are acute angles, we offer *a cute* proof. Draw the perpendicular line segment from B to $\overline{AC}$ (called an *altitude* of triangle ABC), which has length h, intersecting at the point we call X, as shown below.

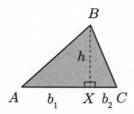

$\overline{AC}$ can be split into two segments $\overline{AX}$ and $\overline{XC}$, which have respective lengths b_1 and b_2, where $b_1 + b_2 = b$. And since BXA and BXC are right triangles, then the previous case tells us that they have respective areas $\frac{1}{2}b_1h$ and $\frac{1}{2}b_2h$. Hence triangle ABC has area

$$\frac{1}{2}b_1h + \frac{1}{2}b_2h = \frac{1}{2}(b_1 + b_2)h = \frac{1}{2}bh$$

as desired.

When angle A or C is obtuse, we get a picture that looks like this.

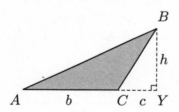

In the acute case, we expressed triangle ABC as the *sum* of two right triangles. Here we express ABC as the *difference* of two right triangles, ABY and CBY. The big right triangle ABY has base length $b + c$, so it has area $\frac{1}{2}(b + c)h$. The smaller right triangle CBY has area $\frac{1}{2}ch$. Hence triangle ABC has area

$$\frac{1}{2}(b + c)h - \frac{1}{2}ch = \frac{1}{2}bh$$

as desired.   ☺

The Pythagorean Theorem

The Pythagorean theorem, perhaps the most famous theorem of geometry and indeed one of the most famous formulas in mathematics, deserves a section of its own. In a right triangle, the side opposite the right angle is called the *hypotenuse*. The other two sides are called *legs*. The right triangle below has legs $\overline{BC}$ and $\overline{AC}$ and hypotenuse $\overline{AB}$, with respective lengths a, b, and c.

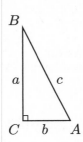

Pythagorean theorem: For a right triangle with leg lengths a and b and hypotenuse length c,

$$a^2 + b^2 = c^2$$

There are reportedly over three hundred proofs of the Pythagorean theorem, but we'll present the simplest ones here. Feel free to skip some of the proofs. My goal is that at least one of the proofs makes you smile or say, "That's pretty cool!"

Proof 1: In the picture on the next page, we have assembled four right triangles to create a giant square.

Question: What is the area of the giant square?

Answer 1: Each side of the square has length $a + b$, so the area is $(a + b)^2 = a^2 + 2ab + b^2$.

Answer 2: On the other hand, the giant square consists of four triangles, each with area $ab/2$, along with a tilted square in the middle with area c^2. (Why is the middle object a square? We know that all four sides are equal and we can use symmetry to see that all four angles are equal: if we rotated the figure 90 degrees, it would be identical, and so each of the middle square's angles must be the same. Since the sum of the angles in a quadrilateral is 360 degrees, we know that each angle must be 90 degrees.) Therefore the area is $4(ab)/2 + c^2 = 2ab + c^2$.

Equating answers 1 and 2 gives us

$$a^2 + 2ab + b^2 = 2ab + c^2$$

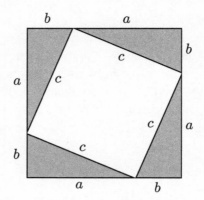

Compute the area of the big square in two different ways.
When you compare your two answers, the Pythagorean theorem pops out.

Subtracting $2ab$ from both sides gives us

$$a^2 + b^2 = c^2$$

as desired.   ☺

Proof 2: Using the same picture as above, we rearrange the triangles as in the figure below. In the first picture the area not occupied by triangles is c^2. In the new picture, the area not occupied by triangles is seen to be $a^2 + b^2$. Thus $c^2 = a^2 + b^2$, as desired.   ☺

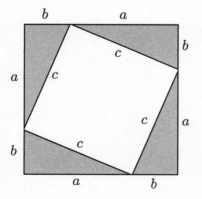 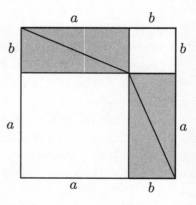

Compare the area of the white space in this figure and the previous one: $a^2 + b^2 = c^2$

Proof 3: This time, let's rearrange the four triangles to form a more compact square on the opposite page with area c^2. (One reason this object is a square is that each corner is composed of $\angle A$ and $\angle B$, which sum to 90°.) As before, the four triangles contribute an area of $4(ab/2) = 2ab$. The tilted square in the middle has area $(a-b)^2 = a^2 - 2ab + b^2$. Hence the combined area equals $2ab + (a^2 - 2ab + b^2) = a^2 + b^2$, as desired.   ☺

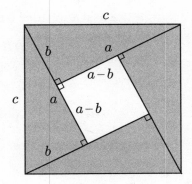

The area of this figure is both c^2 and $a^2 + b^2$

Proof 4: Here's a *similar* proof, and by that I mean, let's exploit what we know about similar triangles. In right triangle ABC, draw the line segment $\overline{CD}$ perpendicular to the hypotenuse, as shown below. Notice

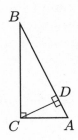

Both of the two smaller triangles are similar to the large one

that the triangle ADC contains both a right angle and $\angle A$, so its third angle must be congruent to $\angle B$. Likewise, triangle CDB has a right angle and $\angle B$, so its third angle must be congruent to $\angle A$. Consequently, all three triangles are similar:

$$\triangle ACB \sim \triangle ADC \sim \triangle CDB$$

Note that the order of the letters matters. We have $\angle ACB = \angle ADC = \angle CDB = 90°$ are all right angles; likewise, $\angle A = \angle BAC = \angle CAD =$

$\angle BCD$, and $\angle B = \angle CBA = \angle DCA = \angle DBC$. Comparing side lengths from the first two triangles gives us

$$AC/AB = AD/AC \Rightarrow AC^2 = AD \times AB$$

Comparing side lengths from the first and third triangles gives us

$$CB/BA = DB/BC \Rightarrow BC^2 = DB \times AB$$

Adding these equations, we have

$$AC^2 + BC^2 = AB \times (AD + DB)$$

And since $AD + DB = AB = c$, we have our desired conclusion:

$$b^2 + a^2 = c^2 \qquad \qquad ☺$$

The next proof is purely geometrical. It uses no algebra, but it does require some visualization skills.

Proof 5: This time we start with two squares, with areas a^2 and b^2, placed side by side, as below. Their combined area is $a^2 + b^2$. We can dissect this object into two right triangles (with side lengths a and b and hypotenuse length c) and a third strange-looking shape. Note that the angle at the bottom of the strange shape must be 90° since it is surrounded by $\angle A$ and $\angle B$. Imagine a hinge placed in the upper left corner of the big square and the upper right corner of the smaller square.

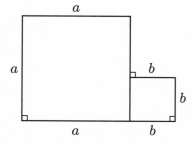 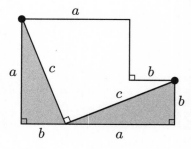

These two squares with area $a^2 + b^2$ can be transformed into ...

Now imagine that the bottom left triangle is "swiveled" 90° in a counter-clockwise direction so it rests on the outside of the top of the large square. Then swivel the other triangle clockwise 90° so that the right angles match up and it sits comfortably in the corner made by the

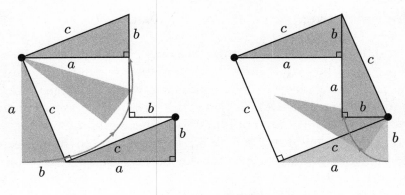

... a square of area c^2!

two squares, as shown in the figure above. The end result is a tilted square of area c^2. Thus $a^2 + b^2 = c^2$, as promised.   ☺

We can apply the Pythagorean theorem to explain question 4 about the football field at the beginning of the chapter with a rope of length 361 feet connecting two goalposts that are 360 feet apart.

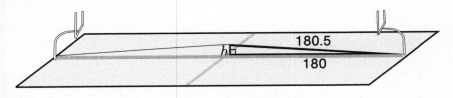

By the Pythagorean theorem, $h^2 + 180^2 = 180.5^2$

The distance from either goalpost to the middle of the field is 180 feet. After the rope is raised to its highest point, h, we create a right triangle, as shown below, with leg length 180 and hypotenuse 180.5. Thus by the Pythagorean theorem, and a few lines of algebra, we get

$$h^2 + 180^2 = 180.5^2$$

$$h^2 + 32,400 = 32,580.25$$

$$h^2 = 180.25$$

$$h = \sqrt{180.25} \approx 13.43 \text{ feet}$$

Hence the rope would be high enough that most trucks could drive under it!

Geometrical Magic

Let's end this chapter, as it began, with a magic trick based on geometry. Most proofs of the Pythagorean theorem involve rearranging the pieces of one geometrical object to obtain another with the same area. But consider the following paradox. Starting with an 8-by-8 square, as pictured below, it looks like we can cut it into four pieces (all of which have Fibonacci number lengths of 3, 5, or 8!), then reassemble those pieces to create a 5-by-13 rectangle. (Try this yourself!) But this should be impossible, since the first figure has area $8 \times 8 = 64$, while the second figure has area $5 \times 13 = 65$. What's going on?

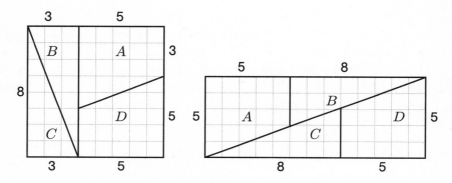

Can a square of area 64 be reassembled to create a rectangle of area 65?

The secret to this paradox is that the diagonal "line" of the 5-by-13 rectangle is not really a straight line. For example, the triangle labeled C has a hypotenuse with slope $3/8 = 0.375$ (since its y-coordinate increases by 3 as its x-coordinate increases by 8), whereas, the top of the figure labeled D (a *trapezoid*) has a slope of $2/5 = 0.4$ (since its y-coordinate increases by 2 as its x-coordinate increases by 5). Since the slopes are different, it doesn't form a straight line, and the same phenomenon occurs with the bottoms of the upper trapezoid and triangle as well. Hence, if we actually looked closely at the rectangle, like in the figure on the opposite page, we'd see a little bit of extra space between the two "almost diagonal lines." And in that space, spread out over a large region, is exactly one extra unit of area.

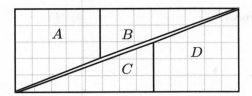

The rectangle contains one unit of extra area spread out across the diagonal

In this chapter, we derived many important properties of triangles, squares, rectangles, and other polygons, all of which are created with straight lines. The investigation of circles and other curved objects will require more sophisticated geometric ideas through trigonometry and calculus, all of which will depend on the magical number π.

CHAPTER EIGHT

3.141592653589...

The Magic of π

Circular Reasoning

We began the last chapter with some problems designed to challenge your geometric intuition pertaining to rectangles and triangles, ending with a problem that involved a rope connecting two goalposts at the opposite ends of a football field. In this chapter our focus will be on circles, and we'll begin with a problem that starts by putting a rope around the Earth!

Question 1. Imagine that a rope is about to be tied around the Earth's equator (approximately 25,000 miles long). Before tying the ends together, an extra 10 feet of rope is added. If we now somehow lift the rope so that it hovers over each point of the equator by the same distance, about how high will the rope be?

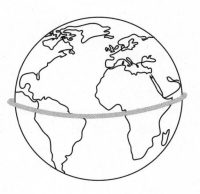

A) Less than one inch off the ground.
B) Just high enough to crawl under.
C) Just high enough to walk under.
D) High enough to drive a truck under.

Question 2. Two points X and Y are fixed on a circle, as in the figure below. We wish to choose a third point Z on the circle somewhere on the major arc (the long arc between X and Y, not the short arc). Where should we choose the point Z to maximize the angle $\angle XZY$?

A) Point A (opposite the midpoint of X and Y).

B) Point B (the reflection of X through the center of the circle).

C) Point C (as close to X as we can).

D) It doesn't matter. All angles will have the same measure.

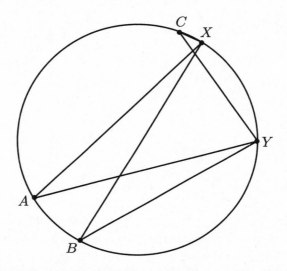

Which point on the major arc between X and Y results in largest angle? Is it angle $\angle XAY$, $\angle XBY$, $\angle XCY$, or are they all the same?

To answer these questions, we need to improve our understanding of circles. (Well, I suppose you don't need circles to read the answers. The answers are B and D, respectively. But in order to appreciate why these answers are true, we need to understand circles.) A circle can be described by a point O and a positive number r so that every point on the circle has a distance r away from O, as shown on the opposite page. The point O is called the *center* of the circle. The distance r is called the *radius* of the circle. As a mathematical convenience, a line segment $\overline{OP}$ from O to a point P on the circle is also called a radius.

Circumference and Area

For any circle, its *diameter D* is defined to be twice its radius, and is the distance across the circle. That is,

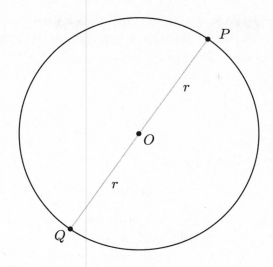

A circle with center O, radius r, and diameter $D = 2r$

$$D = 2r$$

The perimeter of a circle (its distance around) is called the circle's *circumference*, denoted by C. From the picture, it is clear that C is bigger than $2D$, since the distance along the circle from P to Q is bigger than D and the distance from Q back to P is also bigger than D. Consequently, $C > 2D$. If you eyeball it, you might even convince yourself that C is a little bigger than $3D$. (But to see it clearly, you might need to wear 3-D glasses. Sorry.)

Now, if you wanted to go about comparing a circular object's circumference to its diameter, you might wrap a string around the circumference. Then divide the length you measured by the diameter. You'll find, regardless of whether you are measuring a coin, the base of a glass, a dinner plate, or a giant hula hoop, that

$$C/D \approx 3.14$$

We define the number π (pi, pronounced "pie"; the Greek letter for the "p" sound) to be the exact constant that represents the ratio of a circle's circumference to its diameter. That is,

$$\pi = C/D$$

and π is the same for every circle! Or if you prefer, you can write this as a formula for the circumference of any circle. Given the diameter D or radius r of any circle, we have

$$C = \pi D$$

or

$$C = 2\pi r$$

The digits of π begin as follows:

$$\pi = 3.14159\ldots$$

We will provide more digits of π and discuss some of its numerical properties later in this chapter.

> ### ✄Aside
>
> Interestingly, the human eye is not so good at estimating circumferences. For example, take any large drinking glass. What do you think is bigger, its height or its circumference? Most people think the height is bigger, but it's usually the circumference. To convince yourself, put your thumb and middle finger on opposite sides of a glass to determine its diameter. You will likely see that your glass is less than 3 diameters tall.

We can now answer question 1 from the beginning of the chapter. If we think of the equator of the Earth as a perfect circle with circumference $C = 25{,}000$ miles, then its radius must be

$$r = \frac{C}{2\pi} = \frac{25{,}000}{6.28} \approx 4000 \text{ miles}$$

But we don't actually need to know the value of the radius to answer this question. All we really need to know is how much the radius will change if we increase the circumference by 10 feet. Adding 10 feet to the circumference will create a slightly larger circle with a radius that is larger by exactly $10/2\pi = 1.59$ feet. Hence there would be enough space beneath the rope that you could crawl under it (but not walk under it, unless you were quite a limbo dancer!). What is especially surprising about this problem is that the answer of 1.59 feet does not depend at all on the Earth's actual circumference. You would get the same answer with any planet or with a ball of any size! For example, if we have a circle with circumference $C = 50$ feet, then its radius is $50/(2\pi) \approx 7.96$. If we increase the circumference by 10 feet, then the new radius is $60/(2\pi) \approx 9.55$, which is bigger by about 1.59 feet.

✗ Aside

Here's another important fact about circles.

Theorem: Let X and Y be opposite points on a circle. Then for any other point P on the circle, $\angle XPY = 90°$.

For example, in the figure below, angles $\angle XAY$, $\angle XBY$, and $\angle XCY$ are all right angles.

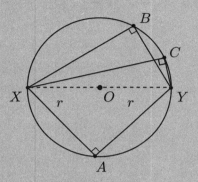

Proof: Draw the radius from O to P, and suppose that $\angle XPO = x$ and $\angle YPO = y$. Our goal is to show that $x + y = 90°$.

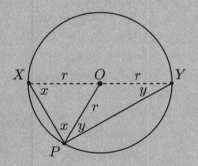

Since $\overline{OX}$ and $\overline{OP}$ are radii of the circle, they both have length r, and therefore triangle XPO is isosceles. By the isosceles triangle theorem, $\angle OXP = \angle XPO = x$. Similarly, $\overline{OY}$ is a radius and $\angle OYP = \angle YPO = y$. Since the angles of triangle XYP must sum to $180°$, we have $2x + 2y = 180°$, and therefore $x + y = 90°$, as desired.   ☺

The theorem above is a special case of one of my favorite theorems from geometry — the central angle theorem, described in the next aside.

✂ **Aside**

The answer to question 2 from the beginning of the chapter is revealed by the *central angle theorem*. Let X and Y be any two points on the circle. The *major arc* is the longer of the two arcs connecting X and Y. The shorter arc is called the *minor arc*. The central angle theorem says that the angle $\angle XPY$ will be the *same* for every point P on the major arc between X and Y. Specifically, angle $\angle XPY$ will be *half* of the *central angle* $\angle XOY$. If Q is on the minor arc from X to Y, then $\angle XQY = 180° - \angle XPY$.

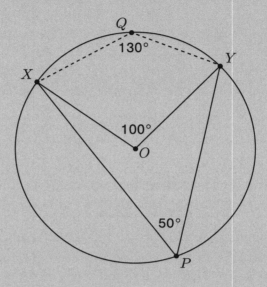

For example, if $\angle XOY = 100°$, then every point P on the major arc from X to Y has $\angle XPY = 50°$, and every point Q on the minor arc from X to Y has $\angle XQY = 130°$.

Once we know the circumference of a circle, we can derive the important formula for the area of a circle.

Theorem: The area of a circle with radius r is πr^2.

This is a formula that you probably had to memorize in school, but it is even more satisfying to understand why it is true. A perfectly rigorous proof requires calculus, but we can give a pretty convincing argument without it.

Proof 1: Think of a circle as consisting of a bunch of concentric rings, as pictured on the following page. Now cut the circle from the top to its center, as shown, then straighten out the rings to form an object that looks like a triangle. What is the area of this triangular object?

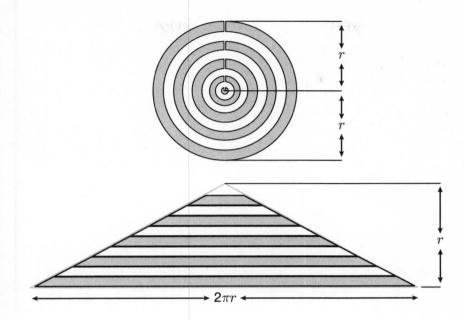

The area of a circle with radius r is πr^2

The area of a triangle with base b and height h is $\frac{1}{2}bh$. For the triangle-like structure the base is $2\pi r$ (the circumference of the circle) and the height is r (the distance from the center of the circle to the bottom). Since the peeled circle becomes more and more triangular as we use more and more rings, then the circle has area

$$\frac{1}{2}bh = \frac{1}{2}(2\pi r)(r) = \pi r^2$$

as desired.   ☺

For a theorem so nice, let's prove it twice! The last proof treated the circle like an onion. This time we treat the circle like a pizza.

Proof 2: Slice the circle into a large number of equally sized slices, then separate the top half from the bottom half and interweave the slices. We illustrate with 8 slices, then with 16 slices, on the next page.

As the number of slices increases, the slices become more and more like triangles with height r. Interlacing the bottom row of triangles (think stalagmites) with the top row of triangles (stalactites) gives us an object that is very nearly a rectangle with height r and base equal to half the circumference, namely πr. (To make it look even more like a rectangle, instead of a parallelogram, we can chop the leftmost stalactite in half and move half of it to the far right.) Since the sliced circle

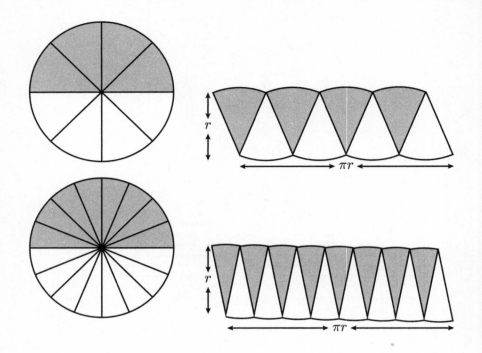

Another proof (by pizza pi?) that the area is πr^2

becomes more and more rectangular as we use more and more slices, the circle has area

$$bh = (\pi r)(r) = \pi r^2$$

as predicted.   ☺

We often want to describe the graph of a circle on the plane. The equation to do so for a circle of radius r centered at $(0,0)$ is

$$x^2 + y^2 = r^2$$

as seen in the graph on the next page. To see why this is true, let (x, y) be any point on the circle, and draw a right triangle with legs of length x and y and hypotenuse r. Then the Pythagorean theorem immediately tells us that $x^2 + y^2 = r^2$.

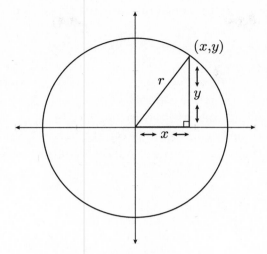

A circle of radius r centered at $(0,0)$ has formula $x^2 + y^2 = r^2$ and area πr^2

When $r = 1$, the above circle is called the *unit circle*. If we "stretch" the unit circle by a factor of a in the horizontal direction and by a factor of b in the vertical direction, then we get an ellipse, like the one below.

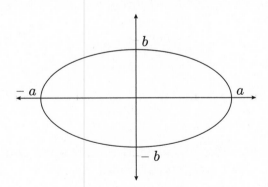

The area of an ellipse is πab

Such an ellipse has the formula

$$\frac{x^2}{a^2} + \frac{y^2}{b^2} = 1$$

and has area πab, which makes sense, because the unit circle has area π and the area has been stretched by ab. Notice that when $a = b = r$, we have a circle of radius r and the $\pi a b$ area formula correctly gives us πr^2.

Here are some fun facts about ellipses. You can create an ellipse by taking two pins, a loop of string, and a pencil. Stick two pins on a piece of paper or cardboard and wrap the string around the pins with a little bit of slack. Place your pencil at one part of the string and pull the string taut so that the string forms a triangle, as in the figure below. Then move the pencil around the two pins, keeping the string taut the entire time. The resulting diagram will be an ellipse.

The positions of the pins are called the *foci* of the ellipse, and they have the following magical property. If you take a marble or billiard ball and place it at one focus, then hit the ball in any direction, after one bounce off the ellipse it will head straight to the other focus.

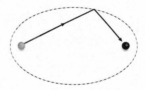

Heavenly bodies like planets and comets travel around the sun in elliptical paths. I can't resist the following rhyme:

Even eclipses
Are based on ellipses!

> ## ⋈ Aside
>
> Interestingly, there is no simple formula for the circumference of an el-
> lipse. But the mathematical genius Srinivasa Ramanujan (1887–1920) es-
> tablished the following excellent approximation. The circumference of
> an ellipse, as described above, is approximately
>
> $$\pi\left(3a + 3b - \sqrt{(3a+b)(3b+a)}\right)$$
>
> Notice that when $a = b = r$, this reduces to $\pi(6r - \sqrt{16r^2}) = 2\pi r$, the
> circumference of a circle.

The number π appears in three-dimensional objects as well. Consider a *cylinder*, such as a can of soup. For a cylinder of radius r and
height h, its *volume* (which measures how much room the shape takes
up) is

$$V_{\text{cylinder}} = \pi r^2 h$$

This formula makes sense, since we can think of the cylinder as made
up of circles with area πr^2 stacked one on top of another (like a stack of
round coasters at a restaurant) to a height of h.

What is the *surface area* of the cylinder? In other words, how much
paint would be necessary to paint its exterior, including the top and
bottom? You do not need to memorize the answer since you can figure
it out by breaking the cylinder into three pieces. The top and bottom
of the cylinder each have an area of πr^2, so they contribute $2\pi r^2$ to the
surface area. For the rest of the cylinder, cut the cylinder with a straight
cut from bottom to top and flatten the resulting object. The object will
be a rectangle with height h and base $2\pi r$, since that's the circumference
of the surrounding circle. Since this rectangle has area $2\pi rh$, the total
surface area of the cylinder is

$$A_{\text{cylinder}} = 2\pi r^2 + 2\pi rh$$

A *sphere* is a three-dimensional object where all points are a fixed
distance away from its center. What is the volume of a sphere of radius
r? Such a sphere would fit inside a cylinder of radius r and height $2r$,
so its volume must be less than $\pi r^2(2r) = 2\pi r^3$. As luck (and calculus)
would have it, the sphere occupies exactly two-thirds of that space. In
other words, the volume of a sphere is

$$V_{\text{sphere}} = \frac{4}{3}\pi r^3$$

The surface area of a sphere has a simple formula that is not so simple to derive:

$$A_{\text{sphere}} = 4\pi r^2$$

Let's end this section with examples of where π appears in ice cream and pizza. Imagine an ice cream cone with height h, and where the circle at the top has radius r. Let s be the *slant height* from the tip of the cone to any point on the circle, as shown below. (We can calculate s from the Pythagorean theorem, since $h^2 + r^2 = s^2$.)

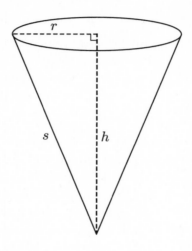

The cone has volume $\pi r^2 h/3$ and surface area $\pi r s$

Such a cone would fit inside a cylinder with radius r and height h, so it is no surprise that the cone's volume is less than $\pi r^2 h$. But it is a surprise (and completely unintuitive without using calculus) that the volume is exactly one-third of that number. In other words,

$$V_{\text{cone}} = \frac{1}{3}\pi r^2 h$$

Although we can derive the surface area without calculus, we just display it for its elegance and simplicity. The surface area of a cone is

$$A_{\text{cone}} = \pi r s$$

Finally, consider a pizza with radius z and thickness a, as shown below. What would its volume be?

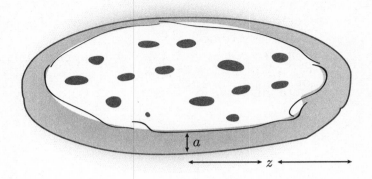

What is the volume of this pizza with radius z and thickness a?

The pizza can be thought of as an unusually shaped cylinder with radius z and height a, so its volume must be

$$V = \pi z^2 a$$

But the answer was really staring you in the face all along, since if we spell out the answer more carefully we get

$$V = pi \, z \, z \, a$$

Some surprising appearances of π

It's no surprise to see π show up in areas and circumferences of circular objects like the ones we have seen. But π shows up in many parts of mathematics where it doesn't seem to belong. Take, for example, the quantity $n!$, which we explored in Chapter 4. There is nothing particularly circular about this number. It is largely used for counting discrete quantities. We know that it's a number that grows extremely fast, and yet there is no efficient shortcut for computing $n!$. For instance, computing 100,000! still requires many thousands of multiplications. And yet there is a useful way to estimate $n!$ using *Stirling's approximation*, which says that

$$n! \approx \left(\frac{n}{e}\right)^n \sqrt{2\pi n}$$

where $e = 2.71828\ldots$ (another important irrational number that we will learn all about in Chapter 10). For instance, a computer can calculate that to four significant figures, $64! = 1.269 \times 10^{89}$. Stirling's approximation says that $64! \approx (64/e)^{64}\sqrt{128\pi} = 1.267 \times 10^{89}$. (Is there a shortcut to raising a number to the 64th power? Yes! Since $64 = 2^6$, you just start with $64/e$, and square it six times.)

The famous *bell curve*, pictured below, which appears throughout statistics and all of the experimental sciences, has a height of $1/\sqrt{2\pi}$. We'll say more about this curve in Chapter 10.

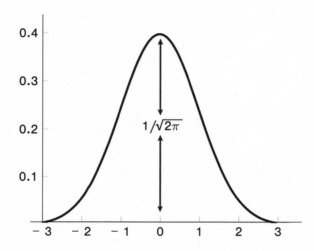

The height of the bell curve is $1/\sqrt{2\pi}$

The number π also often appears in infinite sums. It was Leonhard Euler who first showed that when we add the squares of the reciprocals of the positive integers, we get

$$1 + 1/2^2 + 1/3^2 + 1/4^2 + \cdots = 1 + 1/4 + 1/9 + 1/16 + \cdots = \pi^2/6$$

And if we square each of the above terms, the sum of the reciprocals of fourth powers turns out to be

$$1 + 1/16 + 1/81 + 1/256 + 1/625 + \cdots = \pi^4/90$$

In fact, there are formulas for the sum of reciprocals of every *even* power $2k$, producing an answer of π^{2k}, multiplied by a rational number.

What about powers of odd reciprocals? In Chapter 12, we will show that the sum of the reciprocals of the positive integers is infinite. With odd powers bigger than 1, like the sum of the reciprocals of the cubes,

$$1 + 1/8 + 1/27 + 1/64 + 1/125 + \cdots = ???$$

the sum is finite, but nobody has figured out a simple formula for the sum.

Paradoxically, π pops up in problems pertaining to probability. For example, if you randomly choose two very large numbers, the chance that they have no prime factors in common is a little over 60 percent. More precisely, the probability is $6/\pi^2 = .6079\dots$. It is no coincidence that this is the reciprocal of the answer to one of our earlier infinite sums.

Digits of π

By doing your own careful measurements, you can experimentally determine that π is a little bit bigger than 3, but two questions naturally arise. Can you prove that π is near 3 without the use of physical measurements? And is there a simple fraction or formula for π?

We can answer the first question by drawing a circle of radius 1, which we know has area $\pi 1^2 = \pi$. In the figure below, we have drawn a square with sides of length 2 that completely contains the circle. Since the area of the circle must be less than the area of the square, this proves that $\pi < 4$.

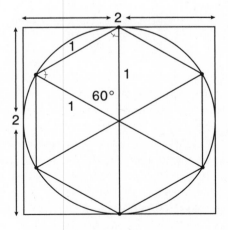

A geometrical proof that $3 < \pi < 4$

On the other hand, the circle also contains a hexagon at six equally spaced points along the circle. What is the *perimeter* of the inscribed hexagon? The hexagon can be broken into 6 triangles, each with a central angle of $360°/6 = 60°$. Two sides of each triangle are radii of the circle with length 1, so the triangle is isosceles. By the isosceles triangle theorem, the other two angles are equal, and must therefore also be $60°$. Hence, these triangles are all equilateral with sides of length 1. The perimeter of the hexagon is 6, which is less than the circumference of 2π. Thus $6 < 2\pi$, and so $\pi > 3$. Putting it all together, we have

$$3 < \pi < 4$$

✂ Aside

We can restrict π to a smaller interval by using polygons with more sides. For example, if we surround the unit circle with a hexagon instead of a square, we can prove that $\pi < 2\sqrt{3} = 3.46\ldots$.

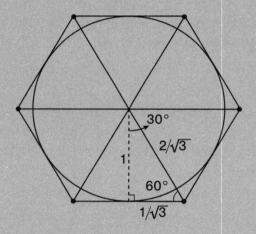

Once again, the hexagon can be subdivided into six equilateral triangles. Each of these triangles can be subdivided into two congruent right triangles. If the short leg has length x, then the hypotenuse has length $2x$, and by the Pythagorean theorem, $x^2 + 1 = (2x)^2$. Solving for x gives us $x = 1/\sqrt{3}$. Consequently, the perimeter of the hexagon is $12/\sqrt{3} = 4\sqrt{3}$, and since this is greater than the circle's circumference, 2π, it follows that $\pi < 2\sqrt{3}$. (Interestingly, we reach the same conclusion by comparing the *area* of the circle to the area of the hexagon.)

The great ancient Greek mathematician Archimedes (287 to 212 BC) built on this result to create inscribed and circumscribed polygons with 12, 24, 48, and 96 sides, leading to $3.14103 < \pi < 3.14271$ and the simpler looking inequality

$$3\frac{10}{71} < \pi < 3\frac{1}{7}$$

There are many simple ways to approximate π as a fraction. For example,

$$\frac{314}{100} = 3.14 \qquad \frac{22}{7} = 3.\overline{142857} \qquad \frac{355}{113} = 3.14159292\ldots$$

I particularly like the last approximation. Not only does it correctly give π to six decimal places, but it also uses the first three odd numbers twice: two 1s, two 3s, and two 5s, in order!

Naturally, it would be interesting to find a fraction that gave us π exactly (where the numerator and denominators are *integers*; otherwise we could simply say $\pi = \frac{\pi}{1}$). In 1768, Johann Heinrich Lambert proved that such a search would be futile, by showing that π is irrational. Perhaps it could be written in terms of square roots or cube roots of simple numbers? For example, $\sqrt{10} = 3.162\ldots$ is pretty close. But in 1882, Ferdinand von Lindemann proved that π is more than just irrational. It is *transcendental*, which means that π is not the root of any polynomial with integer coefficients. For example, $\sqrt{2}$ is irrational, but it is not transcendental, since it is a root of the polynomial $x^2 - 2$.

Although π cannot be expressed as a fraction, it can be expressed as the sum or product of fractions, provided we use infinitely many of them! For example, we will see in Chapter 12 that

$$\pi = 4\left(1 - \frac{1}{3} + \frac{1}{5} - \frac{1}{7} + \frac{1}{9} - \frac{1}{11} + \cdots\right)$$

The formula above is beautiful and startling, yet it is not a very practical formula for calculating π. After 300 terms, we are still not any closer to π than 22/7 is. Here's another astounding formula, called *Wallis's formula*, where π is expressed as an infinite product, although it also takes a long time to converge.

$$\begin{aligned}
\pi &= 4\left(\frac{2}{3} \cdot \frac{4}{3} \cdot \frac{4}{5} \cdot \frac{6}{5} \cdot \frac{6}{7} \cdot \frac{8}{7} \cdot \frac{8}{9} \cdots\right) \\
&= 4\left(1 - \frac{1}{9}\right)\left(1 - \frac{1}{25}\right)\left(1 - \frac{1}{49}\right)\left(1 - \frac{1}{81}\right)\cdots
\end{aligned}$$

Celebrating and Memorizing π (and τ)

Because of people's fascination with π (and as a way to test the speed and accuracy of supercomputers), π has been calculated to trillions of digits. We certainly don't *need* to know π to that level of accuracy. With just forty digits of π, you can measure the circumference of the known universe to within the radius of a hydrogen atom!

The number π has developed almost a cult following. Many people like to celebrate the number π on Pi Day, March 14 (with numeric representation 3/14), which also happens to be the birthdate of Albert Einstein. A typical Pi Day event might consist of mathematically themed pies for display and consumption, Einstein costumes, and of course π memorization contests. Students generally memorize dozens of digits of π, and it is not unusual for the winner to have memorized over a hundred digits. By the way, the current world record for π memorization belongs to Chao Lu of China, who in 2005 recited π to 67,890 decimal places! According to the *Guinness Book of World Records*, Lu practiced for four years to reach this many digits, and it took him a little over twenty-four hours to recite all the digits.

Behold the first 100 digits of π:

$$\pi = 3.1415926535897932384626433832795028841971693993751058209749445923078164062862089986280348253421170671017\ldots$$

Over the years, people have come up with creative ways to memorize the digits of π. One method is to create sentences where the length of each word gives us the next digit of π. Some famous examples include "How I wish I could calculate pi" (which yields seven digits: 3.141592) and "How I want a drink, alcoholic of course, after the heavy lectures involving quantum mechanics" (which provides fifteen digits).

A most impressive example was written in 1995 by Mike Keith, who generated 740 digits in an amazing parody of Edgar Allan Poe's poem "The Raven." The first stanza, along with the title, generates 42 digits. The stanza's "disturbing" ten-letter word generates the digit 0.

Poe, E. Near a Raven

Midnights so dreary, tired and weary.
Silently pondering volumes extolling all by-now obsolete lore.
During my rather long nap—the weirdest tap!
An ominous vibrating sound disturbing my chamber's antedoor.
"This," I whispered quietly, "I ignore."

Keith went on to extend this masterpiece by writing a 3835-digit "Cadaeic Cadenza." (Note that if you replace C with 3, A with 1, D with 4, and so on, then "cadaeic" becomes 3141593.) It begins with the "Raven" parody, but also includes digital commentaries and parodies of other poems such as Lewis Carroll's "Jabberwocky." His most recent contribution to this genre is *Not a Wake: A Dream Embodying π's Digits Fully for 10000 Decimals.* (Note the word lengths in the book's title!)

The word length method for memorizing π suffers from a significant problem. Even if you could memorize the sentences, poems, and stories, it is not so easy to instantly determine the number of letters in each word. Or as I like to say, "How I wish I could elucidate to others. There are often superior mnemonics!" (which yields thirteen digits).

My favorite way to memorize numbers is through the use of a *phonetic code* called the *major system*. In this code, every digit is represented by one or more consonant sounds. Specifically,

1 = t or d
2 = n
3 = m
4 = r
5 = l
6 = j, ch, or sh
7 = k or hard g
8 = f or v
9 = p or b
0 = s or z

There are even mnemonics for memorizing this mnemonic system! My friend Tony Marloshkovips offers the following suggestions. The letter t (or its phonetically similar d) has 1 downstroke; n has 2 downstrokes; m has 3 downstrokes; "four" ends in the letter r; displaying 5 fingers, you see an L between your index finger and your thumb; a backward 6 looks like a j; two 7s can be drawn to form a K; a skater does a $f$igure 8; turning the 9 backward or upside down, you obtain p or b; "zero" begins with z. Or if you prefer, you can put all the consonants in order, TNMRLShKVPS, and you get my (fictitious) friend's name: Tony Marloshkovips.

We can use this code to turn numbers into words by inserting vowel sounds around the associated consonant sounds. For example, the number 31, which uses consonants m and t (or m and d), can be turned into words like

31 = mate, mute, mud, mad, maid, mitt, might, omit, muddy

Notice that a word like "muddy" or "mitt" is acceptable, since the *d* or *t sound* only occurs once. Spelling doesn't matter. Since the consonant sounds of *h*, *w*, and *y* are not represented on the list, then it is acceptable to use those sounds as freely as vowels. Thus, we could also turn 31 into words like "humid" or "midway." Notice that although a number can often be represented by many different words, a word can only be represented by a single number.

The first three digits of π, with consonant sounds *m*, *t*, and *r*, can become words like

314 = meter, motor, metro, mutter, meteor, midyear, amateur

The first five digits 31415, can become "my turtle." Extending this to the first twenty-four digits of π, 314159265358979323846264 can become

My turtle Pancho will, my love, pick up my new mover Ginger

I turn the next seventeen digits 33832795028841971, into

My movie monkey plays in a favorite bucket

I like the next nineteen digits, 6939937510582097494, since they allow some long words:

Ship my puppy Michael to Sullivan's backrubber

The next eighteen digits of π, 459230781640628620, could give us

A really open music video cheers Jenny F. Jones

followed by twenty-two more digits, 8998628034825342117067:

Have a baby fish knife so Marvin will marinate the goose chick!

Thus, with five silly sentences, we have encoded the first one hundred digits of π!

The phonetic code is quite useful for memorizing dates, phone numbers, credit card numbers, and more. Try it, and with a little bit of practice, you will vastly improve your ability to remember numbers.

All mathematicians agree that π is one of the most important numbers in mathematics. But if you look at the formulas and applications that use π, you will find that most of them have π multiplied by 2. The

Greek letter τ (tau, rhymes with "wow") has been adopted to represent this quantity

$$\tau = 2\pi$$

Many people believe that if we could go back in time, mathematical formulas and key concepts in trigonometry would be expressed more simply using τ instead of π. These ideas have been elegantly and entertainingly expressed in articles by Bob Palais ("π is Wrong!") and Michael Hartl ("The Tau Manifesto"). The "central point" of the argument is that circles are defined in terms of their radius, and when we compare the circumference to the radius, we have $C/r = 2\pi = \tau$. Some textbooks are now labeled as "τ-compliant" to indicate that they will express formulas in terms of both π and τ. (Although switching to this constant may not be "easy as pie," many students and teachers will agree that τ is *easier* than π.) It will be interesting to see what happens to this movement in the coming decades. Supporters of τ (who call themselves tauists) earnestly believe that they have the truth on their side, but they are tolerant of the more traditional notation. As they like to say, tauists are never *pious*.

Here are the first one hundred digits of τ with spaces inserted for the mnemonics given afterward. Note that τ begins with the numbers 6 and 28, both of which are *perfect numbers*, as described in Chapter 6. Is that a coincidence? Of course! But it's a fun tidbit anyway.

τ = 6.283185 30717958 64769252 867665 5900576 839433 8798750

211641949 8891846 15632 812572417 99725606 9650684 234135 . . .

In 2012, thirteen-year-old Ethan Brown established a world record by memorizing 2012 digits of τ as a fund-raising project. He used the phonetic code, but instead of creating long sentences, he created visual images, where each image contained a subject, an action (always ending in -ing) and an object being acted upon. The first seven digits, 62 831 85, became "An ocean vomiting a waffle." Here are his images for the first one hundred digits of τ.

An ocean vomiting a waffle
A mask tugging on a bailiff
A shark chopping nylon
Fudge coaching a cello
Elbows selling a couch
Foam burying a mummy
Fog paving glass
A handout shredding a prop

FIFA beautifying the Irish
A doll shooing a minnow
A photon looking neurotic
A puppy acknowledging the sewage
A peach losing its chauffeur
Honey marrying oatmeal

To make the images even easier to remember, Brown adopted the *memory palace* approach by imagining himself wandering through his school, and as he walked down certain passageways and entered various classrooms, there would be three to five objects doing silly things in each room. Ultimately, he had 272 images split up among more than 60 locations. It took him about 4 months of preparation to recite the 2012 digits, which he did in 73 minutes.

Let's end this chapter with a musical celebration of π. I wrote this as a lyrical addition to Larry Lesser's parody, "American Pi." You should only sing the song once, because π doesn't repeat.

A long, long, time ago,
I can still remember how my math class used to make me snore.
'Cause every number we would meet
Would terminate or just repeat,
But maybe there were numbers that did more.

But then my teacher said, "I dare ya
To try to find the circle's area."
Despite my every action,
I couldn't find a fraction.

I can't remember if I cried,
The more I tried or circumscribed,
But something touched me deep inside
The day I learned of pi!

Pi, pi, mathematical pi,
Twice eleven over seven is a mighty fine try.
A good old fraction you may hope to supply,
But the decimal expansion won't die.
Decimal expansion won't die.

Pi, pi, mathematical pi,
3.141592653589.
A good old fraction you may hope to define,
But the decimal expansion won't die!

CHAPTER NINE

$$20° = \pi / 9$$

The Magic of Trigonometry

The High Point of Trigonometry

The subject of trigonometry allows us to solve geometrical problems that can't be solved using classical geometry. For example, consider the following problem.

Question: Using only a protractor and pocket calculator, determine the height of a nearby mountain.

We will provide *five* different methods for solving this problem. The first three methods actually require almost no math whatsoever!

Method 1 (brute force approach): Climb to the top of the mountain and hurl your calculator off the mountain. (This may require considerable force.) Measure the time it takes for your calculator to hit the ground (or listen for the scream of a backpacker below). If the time is t seconds, and if we ignore the effects of air resistance and terminal velocity, then standard physics equations indicate that the height of the mountain is approximately $16t^2$ feet. The disadvantage of this approach is that the effects of air resistance and terminal velocity can be quite significant, so your calculation will be inaccurate. Also, you are unlikely to recover your calculator. And it requires a timekeeping device, which might have been on your calculator. The advantage of this method is that the protractor does not need to be used.

Method 2 (method of tan gents): Find a friendly park ranger and offer her your shiny new protractor if she tells you the height of the mountain. If no park ranger can be found, look for a gentleman with a nice tan who has probably spent considerable time outdoors and might well know the answer to your question. The advantage of this approach is that you may make a new friend and you don't need to surrender your calculator. Also, if you are suspicious of the tan gent's response, you can still climb the mountain and apply method 1. The disadvantage is that you may lose your protractor and be accused of bribery.

Method 3 (law of signs): Before attempting method 1 or 2, look for a posted sign that tells you the height of the mountain. This has the advantage that you don't need to surrender any of your equipment. ☺

Of course, if none of these three methods appeal to you, then we must resort to more mathematical solutions, which are the subject of this chapter.

Trigonometry and Triangles

The word "trigonometry" has Greek roots *trigon* and *metria*, which literally means triangle measurement. We begin with the analysis of some classic triangles.

Isosceles right triangle. In an isosceles right triangle, there is a 90° angle, and the other two angles must be equal. The other angles are thus 45° (since the sum of the angles is 180°), so we refer to such a triangle as a 45-45-90 triangle. If both legs have length 1, then by the Pythagorean theorem, the hypotenuse must have length $\sqrt{1^2 + 1^2} = \sqrt{2}$. Note that all isosceles right triangles will have the same proportions of $1 : 1 : \sqrt{2}$, as pictured below.

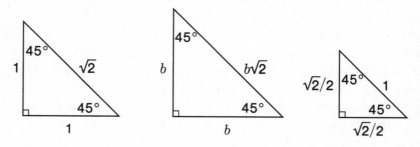

In a 45-45-90 triangle, the side lengths are proportional to $1 : 1 : \sqrt{2}$

30-60-90 triangle. In an *equilateral* triangle, all sides have equal length, and all angles measure 60°. If we divide an equilateral triangle into two congruent halves, as pictured below, we obtain two right triangles with angles measuring 30°, 60°, and 90°. If the sides of the equilateral triangle all have length 2, then the hypotenuse of the right triangle will have length 2, and the short leg will have length 1. By the Pythagorean theorem, the long leg will have height $\sqrt{2^2 - 1^2} = \sqrt{3}$. Thus all 30-60-90 triangles have the same proportions, $1 : \sqrt{3} : 2$ (or as I remember it, they are as easy as $1, 2, \sqrt{3}$). In particular, if the hypotenuse has length 1, then the other side lengths are $1/2$ and $\sqrt{3}/2$.

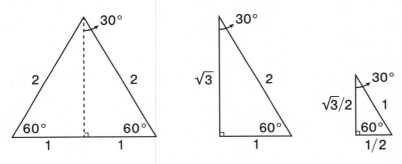

In a 30-60-90 triangle, the sides are proportional to $1 : \sqrt{3} : 2$, respectively

---

✂Aside

When positive integers a, b, c satisfy $a^2 + b^2 = c^2$, we call (a, b, c) a *Pythagorean triple*. The smallest and simplest triple is $(3, 4, 5)$, but there are infinitely many more. Naturally, you could scale your triple by a positive integer to get triples like $(6, 8, 10)$ or $(9, 12, 15)$ or $(300, 400, 500)$, but we would like more interesting examples. Here's a clever way to create Pythagorean triples. Choose *any* two positive numbers m and n where $m > n$. Now let

$$a = m^2 - n^2 \qquad b = 2mn \qquad c = m^2 + n^2$$

Notice that $a^2 + b^2 = (m^2 - n^2)^2 + (2mn)^2 = m^4 + 2m^2n^2 + n^4$, which equals $(m^2 + n^2)^2 = c^2$, so (a, b, c) is a Pythagorean triple. For example, choosing $m = 2, n = 1$ produces $(3, 4, 5)$; $(m, n) = (3, 2)$ gives us $(5, 12, 13)$; $(m, n) = (4, 1)$ yields $(15, 8, 17)$; $(m, n) = (10, 7)$ yields $(51, 140, 149)$. What is especially remarkable (and is proved in any course on number theory) is that *every* Pythagorean triple can be created by this process.

All of trigonometry is based on two important functions: the **sine** and **cosine** functions. Given a right triangle ABC, as pictured below, we let c denote the length of the hypotenuse and let a and b denote the lengths of the sides opposite $\angle A$ and $\angle B$, respectively.

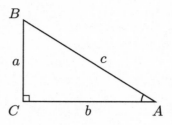

$$\sin A = a/c = \frac{\text{opp}}{\text{hyp}} \qquad \cos A = b/c = \frac{\text{adj}}{\text{hyp}} \qquad \tan A = a/b = \frac{\text{opp}}{\text{adj}}$$

For angle A (which is necessarily acute in a right triangle), we define the *sine of $\angle A$*, denoted $\sin A$, to be

$$\sin A = \frac{a}{c} = \frac{\text{length of leg opposite } A}{\text{length of hypotenuse}} = \frac{\text{opp}}{\text{hyp}}$$

Similarly, we define the *cosine of $\angle A$* to be

$$\cos A = \frac{b}{c} = \frac{\text{length of leg adjacent to } A}{\text{length of hypotenuse}} = \frac{\text{adj}}{\text{hyp}}$$

(Note that *any* right triangle with angle A will be similar to the original triangle and will have sides of proportional length, so the sine and cosine of A will not depend on how large the triangle is.)

After sine and cosine, the next most commonly used function in trigonometry is the **tangent** function. We define the *tangent of $\angle A$* to be

$$\tan A = \frac{\sin A}{\cos A}$$

In terms of the right triangle, we have

$$\tan A = \frac{\sin A}{\cos A} = \frac{a/c}{b/c} = \frac{a}{b} = \frac{\text{length of leg opposite } A}{\text{length of leg adjacent to } A} = \frac{\text{opp}}{\text{adj}}$$

There are many mnemonics for remembering the formulas for sine, cosine, and tangent. The most popular one is "SOH CAH TOA," where SOH reminds us that sine is opposite/hypotenuse, and similarly for

CAH and TOA. My high school teacher used a mnemonic (which assumed that you proceeded in order of sine, then cosine, then tangent) of Oscar Has A Heap Of Apples (OH-AH-OA). My friends modified this to become Olivia Has A Hairy Old Aunt!

For example, in the 3-4-5 triangle below, we have

$$\sin A = \frac{3}{5} \qquad \cos A = \frac{4}{5} \qquad \tan A = \frac{3}{4}$$

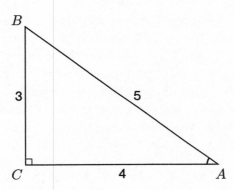

For the 3-4-5 right triangle, $\sin A = 3/5$, $\cos A = 4/5$, $\tan A = 3/4$

Now what about $\angle B$ in the same triangle? If we calculate its sine and cosine values, we see that

$$\sin B = \frac{4}{5} = \cos A \qquad \cos B = \frac{3}{5} = \sin A$$

Here we have $\sin B = \cos A$ and $\cos B = \sin A$. This is not a coincidence, since for any angle $\angle A$, the other acute angle will switch what it considers to be the opposite side and the adjacent side, but it will still have the same hypotenuse. Since $\angle A + \angle B = 90°$, we have for any acute angle

$$\sin(90° - A) = \cos A \qquad \cos(90° - A) = \sin A$$

Thus, for example, if a right triangle ABC has $\angle A = 40°$, then its complement $\angle B = 50°$ has the property $\sin 50° = \cos 40°$ and $\cos 50° = \sin 40°$. In other words, the complement sine is equal to the cosine (which is where the word "cosine" comes from).

There are three other functions that should be part of your trigonometric vocabulary, but they won't be used nearly as much as the first three functions. They are the **secant, cosecant,** and **cotangent** functions, and they are defined as

$$\sec A = \frac{1}{\cos A} \qquad \csc A = \frac{1}{\sin A} \qquad \cot A = \frac{1}{\tan A}$$

You can easily verify that the "co"-functions have the same complementary relationships that sine and cosine do. Namely, for any acute angle in a right triangle, $\sec(90° - A) = \csc A$ and $\tan(90° - A) = \cot A$.

Once you know how to compute the sine of an angle, you use complements to find the cosine of any angle, and from those you can compute tangents and the other trig functions. But how *do* you calculate a sine value, like $\sin 40°$? The simplest way is to just use a calculator. My calculator (in *degree* mode) tells me that $\sin 40° = 0.642\ldots$. How does it do *that* calculation? We'll explain that near the end of this chapter.

There are a handful of trig values that you should know without needing to resort to a calculator. Recall that a 30-60-90 triangle has sides proportional to $1 : \sqrt{3} : 2$, as shown previously. Consequently,

$$\sin 30° = 1/2 \qquad \sin 60° = \sqrt{3}/2$$

and

$$\cos 30° = \sqrt{3}/2 \qquad \cos 60° = 1/2$$

And since a 45-45-90 triangle has sides proportional to $1 : 1 : \sqrt{2}$, we have

$$\sin 45° = \cos 45° = 1/\sqrt{2} = \sqrt{2}/2$$

Since $\tan A = \frac{\sin A}{\cos A}$, I don't think it is necessary to memorize any values of the tangent function except perhaps $\tan 45° = 1$ and that $\tan 90°$ is *undefined* since $\cos 90° = 0$.

Before we determine the height of a mountain using trigonometry, let's first solve the simpler problem of determining the height of a tree. (Would that be using *twig*onometry or *tree*gonometry?)

Suppose you stood 10 feet from a tree, and the angle from the ground to the top of the tree was 50°, as pictured on the opposite page. (By the way, most smartphones have apps that measure angles. With more primitive tools, one can create a functional angle measurer, called a *clinometer*, using a protractor, a straw, and a paper clip.)

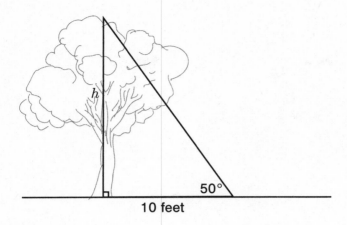

How tall is the tree?

Let h denote the height of the tree. It follows that

$$\tan 50° = \frac{h}{10}$$

and therefore $h = 10 \tan 50°$, which according to the calculator equals $10(1.19\ldots) \approx 11.9$, so the tree is about 11.9 feet tall.

We are now prepared to answer the mountain question by our first mathematical method. The challenge is that we don't know our distance to the center of the mountain. Essentially we have two unknowns (the mountain's height and its distance from us), so we collect two pieces of information. Suppose we measure the angle from our position to the top of the mountain and find the angle to be 40°, then move 1000 feet further away from the mountain and find that the angle now measures 32°, as shown on the next page. Let's use this information to approximate the size of the mountain.

Method 4 (method of tangents): Let h denote the height of the mountain, and let x be our initial distance from the mountain (so x is the length of $\overline{CD}$). Looking at the right triangle BCD, we compute $\tan 40° \approx 0.839$, and therefore

$$\tan 40° \approx 0.839 = \frac{h}{x}$$

which implies $h = 0.839x$. From triangle ABC, we have

$$\tan 32° \approx 0.625 = \frac{h}{x + 1000}$$

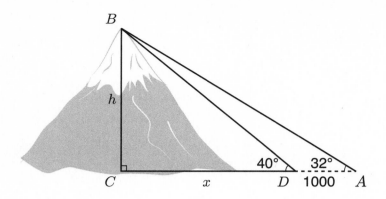

so $h = 0.625(x + 1000) = 0.625x + 625$.

Equating h from both expressions, we get

$$0.839x = 0.625x + 625$$

which has solution $x = 625/(0.214) \approx 2920$. Consequently, h is approximately $0.839(2920) = 2450$, so the mountain is approximately 2450 feet high.

Trigonometry and Circles

So far, we have defined trigonometric functions in terms of a right triangle, and I strongly encourage you to be comfortable with that definition. However, this definition has the shortcoming that it allows us to find the sine, cosine, and tangent only when the angle is strictly between $0°$ and $90°$ (since a right triangle always contains a $90°$ angle and two acute angles). In this section, we define the trigonometric functions in terms of the *unit circle*, which will allow us to find sines, cosines, and tangents for *any* angle whatsoever.

Recall that the unit circle is a circle of radius 1, centered at the *origin* $(0,0)$. It has equation $x^2 + y^2 = 1$, which we derived in the last chapter using the Pythagorean theorem. Suppose I asked you to determine the point (x, y) on the unit circle that corresponds to acute angle A, measured in the counterclockwise direction from the point $(1,0)$, as shown in the next figure.

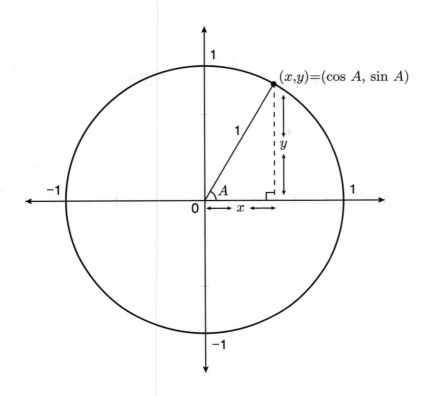

The point (x, y) on the unit circle corresponding to angle A has $x = \cos A$ and $y = \sin A$

We can find x and y by drawing a right triangle and applying our formulas for cosine and sine. Specifically,

$$\cos A = \frac{\text{adj}}{\text{hyp}} = \frac{x}{1} = x$$

and

$$\sin A = \frac{\text{opp}}{\text{hyp}} = \frac{y}{1} = y$$

In other words, the point (x, y) is equal to $(\cos A, \sin A)$. (More generally, if the circle has radius r, then $(x, y) = (r \cos A, r \sin A)$.)

For any angle A, we extend this idea by defining $(\cos A, \sin A)$ to be the point on the unit circle identified with angle A. (In other words, $\cos A$ is the x-coordinate and $\sin A$ is the y-coordinate of the point on the circle with angle A.) Here's the big picture.

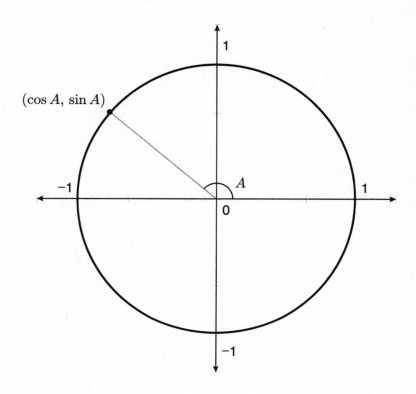

The general definition of cos A and sin A

Here's another big picture, where we have subdivided the unit circle into angles that are 30° apart (with 45° thrown in for good measure), since these correspond to the angles of the special triangles we looked at earlier. We have listed the cosine and sine values for 0°, 30°, 45°, 60°, and 90°. Specifically,

$$(\cos\ 0°, \sin\ 0°) = (1, 0)$$
$$(\cos 30°, \sin 30°) = (\sqrt{3}/2, 1/2)$$
$$(\cos 45°, \sin 45°) = (\sqrt{2}/2, \sqrt{2}/2)$$
$$(\cos 60°, \sin 60°) = (1/2, \sqrt{3}/2)$$
$$(\cos 90°, \sin 90°) = (0, 1)$$

As we'll see, multiples of these angles can be calculated by reflecting the values from the first quadrant.

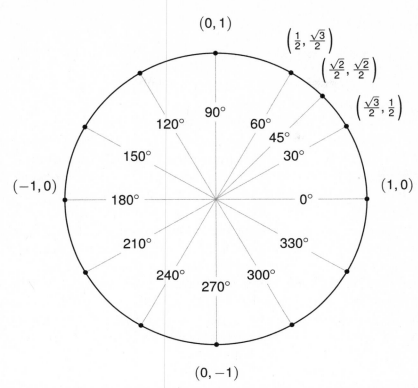

Since adding or subtracting 360° to an angle doesn't really change the angle (it literally takes us full circle), we have for any angle A,

$$\sin(A \pm 360°) = \sin A \qquad \cos(A \pm 360°) = \cos A$$

A negative angle moves in the clockwise direction. For example, the angle $-30°$ is the same as the angle 330°. Notice that when you move A degrees in the clockwise direction, you have the same x-coordinate as when you move A degrees in the counterclockwise direction, but the y-coordinates will have opposite signs. In other words, for any angle A,

$$\cos(-A) = \cos A \qquad \sin(-A) = -\sin A$$

For example,

$$\cos(-30°) = \cos 30° = \sqrt{3}/2 \qquad \sin(-30°) = -\sin 30° = -1/2$$

When we reflect angle A across the y-axis, we get the *supplementary angle* $180 - A$. This keeps the y-value on the unit circle unchanged, but the x-value is negated. In other words,

$$\cos(180 - A) = -\cos A \qquad \sin(180 - A) = \sin A$$

For instance, when $A = 30°$,

$$\cos 150° = -\cos 30° = -\sqrt{3}/2 \qquad \sin 150° = \sin 30° = 1/2$$

We continue to define the other trigonometric functions as before, for instance, $\tan A = \sin A / \cos A$.

The x-axis and y-axis divide the plane into four *quadrants*. We call these quadrants I, II, III, and IV, where quadrant I has angles between $0°$ and $90°$; quadrant II has angles between $90°$ and $180°$; quadrant III has angles between $180°$ and $270°$; and quadrant IV has angles between $270°$ and $360°$. Note that the sine is positive in quadrants I and II, the cosine is positive in quadrants I and IV, and therefore the tangent is positive in quadrants I and III. Some students use the mnemonic All Students Take Calculus (A, S, T, C) to remember which of the trig functions are positive in each respective quadrant (all, sine, tangent, cosine).

The last bit of vocabulary worth learning involves the *inverse trigonometric functions*, which are useful for determining unknown angles. For example, the inverse sine of $1/2$, denoted as $\sin^{-1}(1/2)$, tells us the angle A for which $\sin A = 1/2$. We know that $\sin 30° = 1/2$, so

$$\sin^{-1}(1/2) = 30°$$

The $\sin^{-1}$ function (also called the *arc sine* function) always gives an angle between $-90°$ and $90°$, but be aware that there are other angles outside of this interval that have the same sine value. For example, $\sin 150° = 1/2$, as is any multiple of $360°$ added to $30°$ or $150°$.

For the 3-4-5 triangle pictured on the opposite page, our calculator can determine angle A in three different ways through inverse trig functions:

$$\angle A = \sin^{-1}(3/5) = \cos^{-1}(4/5) = \tan^{-1}(3/4) \approx 36.87° \approx 37°$$

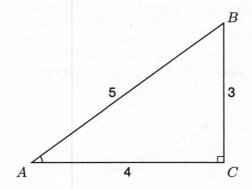

Inverse trig functions can determine angles from side lengths.
Here, since $\tan A = 3/4$, $\angle A = \tan^{-1}(3/4) \approx 37°$.

It's time to put these trigonometric functions to work. In geometry, the Pythagorean theorem tells us the length of the hypotenuse given the lengths of the legs of any right triangle. In trigonometry, we can perform a similar calculation for *any* triangle using the *law of cosines*.

Theorem (law of cosines): For any triangle ABC, where the sides of length a and b form $\angle C$, the third side of length c satisfies

$$c^2 = a^2 + b^2 - 2ab \cos C$$

For example, in the triangle below, triangle ABC has sides of length 21 and 26 with a 15° angle between them. So according to the law of cosines, the third side of length c must satisfy

$$c^2 = 21^2 + 26^2 - 2(21)(26) \cos 15°$$

and since $\cos 15° \approx 0.9659$, this equation reduces to $c^2 = 62.21$, and therefore $c \approx 7.89$.

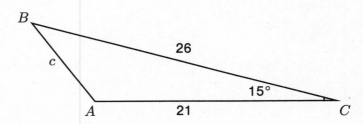

✕ Aside

Proof: To prove the law of cosines, we consider three cases, depending on whether $\angle C$ is a right angle, acute, or obtuse. If $\angle C$ is a right angle, then $\cos C = \cos 90° = 0$, so the law of cosines simply says that $c^2 = a^2 + b^2$, which is true by the Pythagorean theorem.

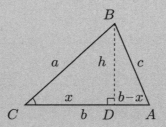

If $\angle C$ is acute, as in the figure above, draw the perpendicular line from B to $\overline{AC}$ intersecting at point D; this splits ABC into two right triangles. From the picture above and the Pythagorean theorem applied to CBD, we have $a^2 = h^2 + x^2$, and therefore

$$h^2 = a^2 - x^2$$

From triangle ABD, we have $c^2 = h^2 + (b - x)^2 = h^2 + b^2 - 2bx + x^2$, and therefore

$$h^2 = c^2 - b^2 + 2bx - x^2$$

Setting the above values of h^2 equal to each other gives us

$$c^2 - b^2 + 2bx - x^2 = a^2 - x^2$$

and therefore

$$c^2 = a^2 + b^2 - 2bx$$

And from right triangle CBD, we see that $\cos C = x/a$, so $x = a \cos C$. Thus, when $\angle C$ is acute,

$$c^2 = a^2 + b^2 - 2ab \cos C$$

If $\angle C$ is obtuse, then we create the right triangle CBD on the outside of the triangle, as shown in the figure opposite.

(continues on the following page)

⚔ Aside (*continued*)

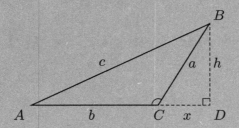

From right triangles CBD and ABD, Pythagoras tells us that $a^2 = h^2 + x^2$ and $c^2 = h^2 + (b+x)^2$. This time, when we equate the values h^2, we get

$$c^2 = a^2 + b^2 + 2bx$$

This time, triangle CBD tells us that $\cos(180° - C) = x/a$, so $x = a\cos(180° - C) = -a\cos C$. So once again, we get the desired equation

$$c^2 = a^2 + b^2 - 2ab\cos C \qquad ☺$$

By the way, there is also a nice formula for the area of the previous triangle.

Corollary: For any triangle ABC, where the sides of length a and b form $\angle C$,

$$\text{area of triangle } ABC = \frac{1}{2}ab\sin C$$

⚔ Aside

Proof: The area of a triangle with base b and height h is $\frac{1}{2}bh$. In all three cases covered in the law of cosines proof, the triangle has a base b; now let's determine h. In the acute case, observe that $\sin C = h/a$, so $h = a\sin C$. In the obtuse case, we have $\sin(180° - C) = h/a$, so $h = a\sin(180° - C) = a\sin C$, as before. In the right-angled case, $h = a$, which equals $a\sin C$, since $C = 90°$ and $\sin 90° = 1$. Thus, since $h = a\sin C$ in all three cases, the area of the triangle is $\frac{1}{2}ab\sin C$, as desired. □

As a consequence of this corollary, notice that

$$\sin C = \frac{2(\text{area of triangle } ABC)}{ab}$$

and therefore

$$\frac{\sin C}{c} = \frac{2(\text{area of triangle } ABC)}{abc}$$

In other words, for the triangle ABC, $(\sin C)/c$ is twice the area of ABC divided by the product of the lengths of all of its sides. But there was nothing special about angle C in this statement. We would get the same conclusion from $(\sin B)/b$ or $(\sin A)/a$. Consequently, we have just proved the following very useful theorem.

Theorem (law of sines): For any triangle ABC with respective side lengths a, b, and c,

$$\frac{\sin A}{a} = \frac{\sin B}{b} = \frac{\sin C}{c}$$

or equivalently,

$$\frac{a}{\sin A} = \frac{b}{\sin B} = \frac{c}{\sin C}$$

We can apply the law of sines to determine the height of the mountain a different way. This time, we focus on a, our original distance to the top of the mountain, as in the illustration below.

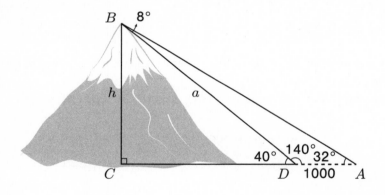

Finding the mountain's height with law of sines

Method 5 (law of sines): In triangle ABD, $\angle BAD = 32°$, $\angle BDA = 180° - 40° = 140°$, and therefore $\angle ABD = 8°$. Applying the law of sines to this triangle, we have

$$\frac{a}{\sin 32°} = \frac{1000}{\sin 8°}$$

Multiplying both sides by $\sin 32°$, we get $a = 1000 \sin 32° / \sin 8° \approx 3808$ feet. Next, since $\sin 40 \approx 0.6428 = h/a$, it follows that

$$h = a \sin 40 \approx (3808)(.6428) = 2448$$

so the mountain is about 2450 feet high, which is consistent with our previous answer.

> ✂ **Aside**
>
> Here's another pretty formula worth knowing, called *Hero's formula*, which tells us the area of a triangle from its side lengths a, b, and c. The formula is straightforward once you calculate the *semi-perimeter*
>
> $$s = \frac{a+b+c}{2}$$
>
> Hero's formula says that the area of the triangle with side lengths a, b, c is
>
> $$\sqrt{s(s-a)(s-b)(s-c)}$$
>
> For example, a triangle with side lengths 3, 14, 15 (first five digits of π) would have $s = (3 + 14 + 15)/2 = 16$. Therefore the triangle has area $\sqrt{16(16-3)(16-14)(16-15)} = \sqrt{416} \approx 20.4$.
>
> Hero's formula can be derived from the law of cosines and a little bit of heroic algebra.

Trigonometric Identities

Trigonometric functions satisfy many interesting relationships, called *identities*. We have seen a few of them already, like

$$\sin(-A) = -\sin A \qquad \cos(-A) = \cos A$$

but there are other interesting identities leading to useful formulas, which we will explore in this section. The first identity comes from the formula for the unit circle:

$$x^2 + y^2 = 1$$

Since the point $(\cos A, \sin A)$ is on the unit circle, it must satisfy that relationship, and therefore $(\cos A)^2 + (\sin A)^2 = 1$. This gives us perhaps the most important identity in all of trigonometry.

Theorem: For any angle A,

$$\cos^2 A + \sin^2 A = 1$$

So far we have mainly been using the letter A to represent an arbitrary angle, but there is certainly nothing special about that letter. The identity above is often stated with other letters. For instance,

$$\cos^2 x + \sin^2 x = 1$$

The Greek letter θ (theta) is another popular choice

$$\cos^2 \theta + \sin^2 \theta = 1$$

And sometimes we simply refer to the identity without any variable at all. For instance, we might abbreviate the theorem as

$$\cos^2 + \sin^2 = 1$$

Before proving the other identities, let's apply the Pythagorean theorem to compute the length of a line segment. This will be key to the proof of our first identity, and it's a useful result by itself.

Theorem (distance formula): Let L be the length of the line segment from (x_1, y_1) to (x_2, y_2). Then

$$L = \sqrt{(x_2 - x_1)^2 + (y_2 - y_1)^2}$$

For example, the length of the line segment from $(-2, 3)$ to $(5, 8)$ would be $\sqrt{(5 - (-2))^2 + (8 - 3)^2} = \sqrt{7^2 + 5^2} = \sqrt{74} \approx 8.6$.

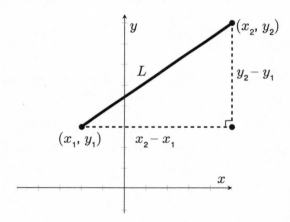

From the Pythagorean theorem, $L^2 = (x_2 - x_1)^2 + (y_2 - y_1)^2$

Proof: Consider two points (x_1, y_1) and (x_2, y_2) as in the figure above. Draw a right triangle so that the line segment connecting them is the hypotenuse of a right triangle. In our picture, the length of the base is $x_2 - x_1$, and the height is $y_2 - y_1$. Hence by the Pythagorean theorem, the hypotenuse L satisfies

$$L^2 = (x_2 - x_1)^2 + (y_2 - y_1)^2$$

and therefore $L = \sqrt{(x_2 - x_1)^2 + (y_2 - y_1)^2}$, as desired.   □

Note that the formula works even when $x_2 < x_1$ or $y_2 < y_1$. For instance, when $x_1 = 5$ and $x_2 = 1$, the distance between x_1 and x_2 is 4, and even though $x_2 - x_1 = -4$, the square of that number is 16, which is all that matters.

✂Aside

In a box of dimensions $a \times b \times c$, what is the length of the diagonal? Let O and P be diagonally opposite corners of the base of the box. The base is an $a \times b$ rectangle, so the diagonal $\overline{OP}$ has length $\sqrt{a^2 + b^2}$.

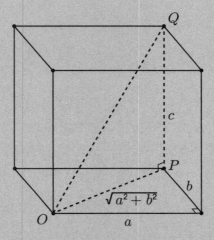

Now if we go straight up a distance c from P, we reach the point Q that is in the opposite corner from O. To find the distance from O to Q, notice that triangle OPQ is a right triangle with leg lengths $\sqrt{a^2 + b^2}$ and c. Hence, by the Pythagorean theorem, the length of the diagonal $\overline{OQ}$ is

$$\sqrt{\sqrt{a^2 + b^2}^2 + c^2} = \sqrt{a^2 + b^2 + c^2}$$

We are now ready to prove a trigonometric identity that is both elegant and useful. The proof of this theorem is a little tricky, so feel free to skip it, but the good news is that once we have done the hard work to establish it, then many more identities will immediately follow.

Theorem: For any angles A and B,

$$\cos(A - B) = \cos A \cos B + \sin A \sin B$$

Proof: On the unit circle centered at O pictured on the next page, let P be the point $(\cos A, \sin A)$ and Q be the point $(\cos B, \sin B)$. Suppose we let c denote the length of $\overline{PQ}$. What can we say about c?

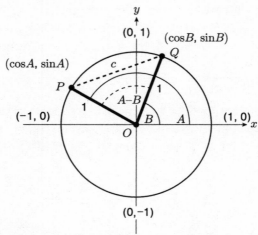

This picture can be used to prove $\cos(A - B) = \cos A \cos B + \sin A \sin B$

In triangle OPQ, we see that $\overline{OP}$ and $\overline{OQ}$ are both radii of the unit circle, so they have length 1 and the angle $\angle POQ$ between them has measure $A - B$. Therefore, by the law of cosines,

$$c^2 = 1^2 + 1^2 - 2(1)(1)\cos(A - B)$$
$$= 2 - 2\cos(A - B)$$

On the other hand, from the distance formula, c satisfies

$$c^2 = (x_2 - x_1)^2 + (y_2 - y_1)^2$$

so the distance c from point $P = (\cos A, \sin A)$ to point $Q = (\cos B, \sin B)$ satisfies

$$c^2 = (\cos B - \cos A)^2 + (\sin B - \sin A)^2$$
$$= \cos^2 B - 2\cos A \cos B + \cos^2 A + \sin^2 B - 2\sin A \sin B + \sin^2 A$$
$$= 2 - 2\cos A \cos B - 2\sin A \sin B$$

where the last line used $\cos^2 B + \sin^2 B = 1$ and $\cos^2 A + \sin^2 A = 1$.
Equating the two expressions for c^2 tells us

$$2 - 2\cos(A - B) = 2 - 2\cos A \cos B - 2\sin A \sin B$$

Subtracting 2 from both sides, then dividing by -2 gives us

$$\cos(A - B) = \cos A \cos B + \sin A \sin B \qquad \square$$

✂ Aside

The proof of the $\cos(A - B)$ formula relied on the law of cosines and assumed that $0° < A - B < 180°$. But we can also prove the theorem without making those assumptions. If we rotate the previous triangle POQ clockwise B degrees, we obtain the congruent triangle $P'OQ'$ where Q' is on the x-axis at the point $(1, 0)$.

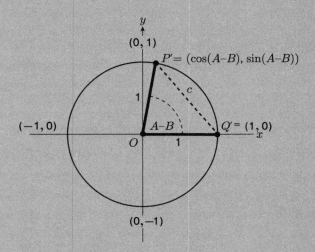

Since $\angle P'OQ' = A - B$, we have $P' = (\cos(A - B), \sin(A - B))$. Thus if we apply the distance formula to $\overline{P'Q'}$, we have

$$c^2 = (\cos(A - B) - 1)^2 + (\sin(A - B) - 0)^2$$
$$= \cos^2(A - B) - 2\cos(A - B) + 1 + \sin^2(A - B)$$
$$= 2 - 2\cos(A - B)$$

so we can conclude that $c^2 = 2 - 2\cos(A - B)$ without using the law of cosines or making any assumptions about the angle $A - B$. The rest of the proof follows as before.

Notice that when $A = 90°$, the $\cos(A - B)$ formula says

$$\cos(90° - B) = \cos 90° \cos B + \sin 90° \sin B$$
$$= \sin B$$

since $\cos 90° = 0$ and $\sin 90° = 1$. If we replace B with $90° - B$ in the above equation, we get

$$\cos B = \cos 90° \cos(90° - B) + \sin 90° \sin(90° - B)$$
$$= \sin(90° - B)$$

Earlier, we saw that these two statements were true when B is an acute angle, but the algebra above makes it true for all angles B. Likewise, if we replace B with $-B$ in the $\cos(A - B)$ theorem, we obtain

$$\cos(A + B) = \cos A \cos(-B) + \sin A \sin(-B)$$
$$= \cos A \cos B - \sin A \sin B$$

since $\cos(-B) = \cos B$ and $\sin(-B) = -\sin B$. When we let $B = A$ above, we get the *double angle* formula:

$$\cos(2A) = \cos^2 A - \sin^2 A$$

and since $\cos^2 A = 1 - \sin^2 A$ and $\sin^2 A = 1 - \cos^2 A$, we also get

$$\cos(2A) = 1 - 2 \sin^2 A \text{ and } \cos(2A) = 2 \cos^2 A - 1$$

We can build on these cosine identities to get related sine identities. For example,

$$\sin(A + B) = \cos(90 - (A + B)) = \cos((90 - A) - B)$$
$$= \cos(90 - A) \cos B + \sin(90 - A) \sin B$$
$$= \sin A \cos B + \cos A \sin B$$

Setting $B = A$ gives us a double angle formula for sines, namely

$$\sin(2A) = 2 \sin A \cos A$$

Or replacing B with $-B$, we have

$$\sin(A - B) = \sin A \cos B - \cos A \sin B$$

Let's summarize many of the identities that we have learned in this chapter so far.

Pythagorean theorem:	$\cos^2 A + \sin^2 A = 1$
Negative angles:	$\cos(-A) = \cos(360° - A) = \cos A$
	$\sin(-A) = \sin(360° - A) = -\sin A$
Supplementary angles:	$\cos(180° - A) = -\cos(A)$
	$\sin(180° - A) = \sin(A)$
Complementary angles:	$\cos(90° - A) = \sin(A)$
	$\sin(90° - A) = \cos(A)$
Cosine of difference:	$\cos(A - B) = \cos A \cos B + \sin A \sin B$
Cosine of sum:	$\cos(A + B) = \cos A \cos B - \sin A \sin B$
Sine of sum:	$\sin(A + B) = \sin A \cos B + \cos A \sin B$
Sine of difference:	$\sin(A - B) = \sin A \cos B - \cos A \sin B$
Double angle formulas:	$\cos(2A) = \cos^2 A - \sin^2 A$
	$\cos(2A) = 1 - 2\sin^2 A$
	$\cos(2A) = 2\cos^2 A - 1$
	$\sin(2A) = 2\sin A \cos A$
For triangle ABC:	Area = $\frac{1}{2}ab \sin C$
Law of cosines:	$c^2 = a^2 + b^2 - 2ab \cos C$
Law of sines:	$\frac{\sin A}{a} = \frac{\sin B}{b} = \frac{\sin C}{c}$

Some useful trigonometric identities

Again, I should point out that although we have written the identities using angle A or B, there is nothing particularly special about these letters. You might well see them written with other angles. For instance, $\cos(2u) = \cos^2 u - \sin^2 u$ or $\sin(2\theta) = 2\sin\theta \cos\theta$.

Radians and Trigonometric Graphs

So far in our discussion of geometry and trigonometry, we have assigned our angles a measure that ranges from 0 to 360 *degrees*. But if you look at the unit circle, there is nothing particularly natural about the number 360. This number was chosen by the ancient Babylonians, probably since they used a base 60 number system and it is approximately the number of days in a year. Instead, for most areas of science and mathematics, it is preferable to measure angles using *radians*. We define

$$2\pi \text{ radians } = 360°$$

Or equivalently,

$$1 \text{ radian } = \frac{180°}{\pi}$$

Or for the tauists out there who like $\tau = 2\pi$,

$$1 \text{ radian } = \frac{360°}{2\pi} = \frac{360°}{\tau}$$

Numerically, 1 radian is approximately 57°. Why are radians more natural than degrees? On a circle of radius r, an angle of 2π radians captures the circle's circumference of $2\pi r$. If we take any fraction of that angle, then the amount of arc that we capture is $2\pi r$ times that fraction. Specifically, 1 radian captures an arc of length $2\pi r(1/2\pi) = r$ and m radians would capture an arc of length mr. In summary, on the unit circle, the angle in radians is equal to its corresponding arc length. How convenient!

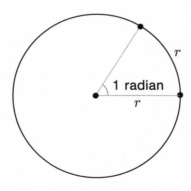

A circle has 2π radians

Here is the unit circle with some common angles given in radians.

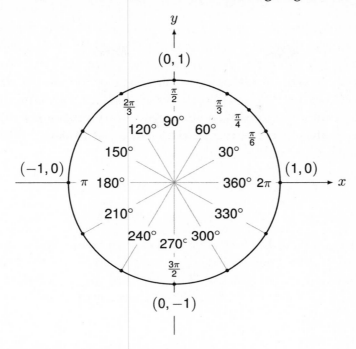

And here's a τ version for comparison.

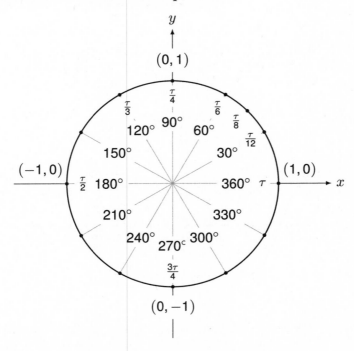

You can see from the pictures one of the reasons some mathematicians prefer τ over π. For a 90° angle, which is one-quarter of the circle, the radian measure is $\tau/4$. For 120°, which is one-third of the circle, the measure is $\tau/3$. Indeed, the letter τ was chosen since it suggested the word *turn*. For instance, 360° is one turn of the circle and has radian measure τ; 60° is one-sixth of a turn and has radian measure $\tau/6$.

As we'll see later in this book, the formulas for computing trigonometric functions are much cleaner when using radians instead of degrees. For example, we can compute sines and cosines as "infinitely long polynomials" with the formulas

$$\sin x = x - x^3/3! + x^5/5! - x^7/7! + x^9/9! - \cdots$$

$$\cos x = 1 - x^2/2! + x^4/4! - x^6/6! + x^8/8! - \cdots$$

but these formula only work when x is expressed in radians. Likewise, in calculus, we will see that the *derivative* of $\sin x$ is $\cos x$, but that is only true when x is in radians. The *graphs* of trigonometric functions $y = \sin x$ and $y = \cos x$ are often given when x is measured in radians.

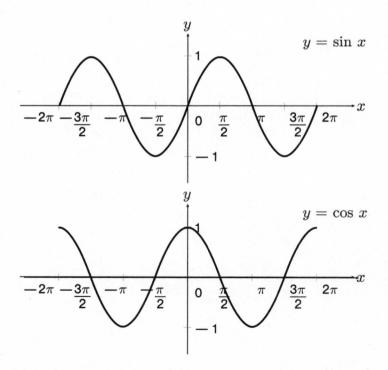

The graphs of $\sin x$ and $\cos x$, where the x variable is measured in radians

Because of the circular nature of sines and cosines, both graphs repeat themselves every 2π units. (Score another point for the tau-ists!) This makes sense because the angle $x + 2\pi$ is the same as the angle x. We say that these graphs have *period* 2π. Moreover, if you shift the cosine graph to the right by $\pi/2$ units, it completely coincides with the sine graph. That's because $\pi/2$ radians is $90°$, and therefore

$$\sin x = \cos(\pi/2 - x)$$
$$= \cos(x - \pi/2)$$

For example, $\sin 0 = 0 = \cos(-\pi/2)$ and $\sin \pi/2 = 1 = \cos 0$.

Since $\tan x = \sin x / \cos x$, it is undefined whenever $\cos x = 0$ (which happens halfway between each multiple of π). The graph of the tangent function has period π, as shown below.

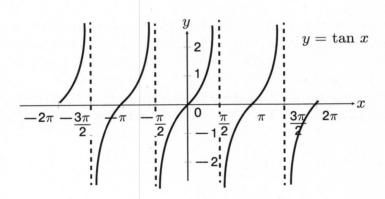

The graph of $y = \tan x$

You can combine sine functions and cosine functions to create almost any function that behaves in a periodic way. This is why trigonometric functions are instrumental in modeling seasonal behavior like temperatures and economic data, or physical phenomena such as sound and water waves, electricity, or the beating of your heart.

Let's end with a magical connection between trigonometry and π. On a calculator, type as many 5s as you can. My calculator allows 5,555,555,555,555,555. Now take the reciprocal of this number. On my calculator, I get

$$1/5{,}555{,}555{,}555{,}555{,}555 = 1.8 \times 10^{-16}$$

Next press the sin button on your calculator (in degree mode) and look at the leading digits (ignoring any initial strings of zeros that may appear). The answer on my display is

$$3.1415926535898 \times 10^{-18}$$

which (after a decimal point, followed by seventeen 0s) are the digits of π to several places! In fact, you should get a similar pi-culiar result if you start with any number of 5s — as long as you have at least five of them.

In this chapter, we have seen how trigonometry helps us better understand triangles and circles. Trigonometric functions interact with each other in many beautiful ways, and we have seen how they are intimately connected with the number π. In the next chapter, we will see that they are also intertwined with two other fundamental numbers, the irrational number $e = 2.71828\ldots$ and the imaginary number i.

CHAPTER TEN

$$e^{i\pi} + 1 = 0$$

The Magic of i and e

The Most Beautiful Mathematical Formula

Every once in a while, mathematics and science journals survey their readers to choose the most beautiful mathematical equations. Inevitably, at the top of this list is the following formula, attributed to Leonhard Euler:

$$e^{i\pi} + 1 = 0$$

Sometimes people refer to this as "God's equation" because it uses perhaps the five most important numbers in mathematics: 0 and 1, which are the foundations of arithmetic; π, the most important number in geometry; e, the most important number in calculus; and i, which may be the most important number in algebra. Even more than that, it uses the fundamental operations of addition, multiplication, and exponentiation. Although we have a good idea about the meaning of 0, 1, and π, it is the goal of this chapter to explore the irrational number e and the imaginary number i so that when we are finished, the formula will be almost as obvious to us as $1 + 1 = 2$ (or at least as easy as $\cos 180° = -1$).

> ✂ **Aside**
>
> Here are some other mathematical equations that were also contenders for being most beautiful. Most of these formulas appear in this book; some have already been discussed, while others are still to come! The first two of these formulas were also discovered by Leonhard Euler.
>
> 1. In any polyhedron (a solid figure made up of flat faces, straight-line edges, and sharp corners called vertices) with V vertices, E edges, and F faces,
> $$V - E + F = 2$$
>   For example, a cube has 8 vertices, 12 edges, and 6 faces, and it satisfies $V - E + F = 8 - 12 + 6 = 2$.
>
> 2. $$1 + 1/4 + 1/9 + 1/16 + 1/25 + \cdots = \pi^2/6$$
>
> 3. $$1 + 1/2 + 1/3 + 1/4 + 1/5 + \cdots = \infty$$
>
> 4. $$0.99999\ldots = 1$$
>
> 5. Stirling's approximation for $n!$:
> $$n! \approx \left(\frac{n}{e}\right)^n \sqrt{2\pi n}$$
>
> 6. Binet's formula for the nth Fibonacci number:
> $$F_n = \frac{1}{\sqrt{5}} \left[\left(\frac{1 + \sqrt{5}}{2}\right)^n - \left(\frac{1 - \sqrt{5}}{2}\right)^n \right]$$

The Imaginary Number i: The Square Root of -1

The number i has the mysterious property that

$$i^2 = -1$$

When people first hear that, they tend to think that it's impossible. How can a number times itself be negative? After all, $0^2 = 0$ and a negative number times itself must be positive. But before you totally dismiss the idea, it is possible that there was a time in your life when you thought that negative numbers were impossible (as most mathematicians did for centuries). What does it mean for a number to be less than 0? How can something be *less than nothing*? Eventually, you came to view numbers as occupants of the *real line* shown opposite, with positive numbers to the right of 0 and negative numbers to the left of 0. In a similar way, we'll need to think outside the box (or outside the line, anyway) to appreciate i, but once we do, we will find that it has a very *real* significance.

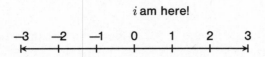

The real line does not contain imaginary numbers. Where could they be hiding?

We call i an *imaginary number*. An imaginary number is any number whose square is a negative number. For example, the imaginary number $2i$ satisfies $(2i)(2i) = 4i^2 = -4$. Algebra with imaginary numbers works just like it does with real numbers. For instance,

$$3i + 2i = 5i, \qquad 3i - 2i = 1i = i, \qquad 2i - 3i = -1i = -i$$

and

$$3i \times 2i = 6i^2 = -6, \qquad \frac{3i}{2i} = 3/2$$

By the way, note that the number $-i$ also has a square of -1, since $(-i)(-i) = i^2 = -1$. Multiplying a real number by an imaginary number has predictable results. For example, $3 \times 2i = 6i$.

What happens when you add a real number and an imaginary number? For instance, what is 3 plus $4i$? The answer is just that: $3 + 4i$. It doesn't simplify further (in the same way that we don't simplify $1 + \sqrt{3}$). Numbers of the form $a + bi$ (where a and b are real numbers) are called *complex* numbers. Note that real and imaginary numbers can be considered special cases of complex numbers (where $b = 0$ and $a = 0$, respectively). Thus the real number π and the imaginary number $7i$ are also complex.

Let's do some examples of (not so) complex arithmetic, beginning with addition and subtraction:

$$(3 + 4i) + (2 + 5i) = 5 + 9i$$

$$(3 + 4i) - (2 + 5i) = 1 - i$$

For multiplication we use the FOIL rule from algebra in Chapter 2:

$$(3+4i)(2+5i) = 6+15i+8i+20i^2$$
$$= (6-20)+(15+8)i$$
$$= -14+23i$$

With complex numbers, every quadratic polynomial $ax^2 + bx + c$ has two roots (or one repeated root). From the quadratic formula, the polynomial will equal 0 whenever

$$x = \frac{-b \pm \sqrt{b^2 - 4ac}}{2a}$$

In Chapter 2, we said there were no real solutions if the number under the square root was negative, but now negative square roots don't bother us. For example, the equation $x^2 + 2x + 5$ has roots

$$x = \frac{-2 \pm \sqrt{4-20}}{2} = \frac{-2 \pm \sqrt{-16}}{2} = \frac{-2 \pm 4i}{2} = -1 \pm 2i$$

By the way, the quadratic formula still works even when a, b, or c is complex.

Quadratic polynomials always have at least one root, although it may be complex. The next theorem says that this is true for almost all polynomials.

Theorem (Fundamental theorem of algebra): Every polynomial $p(x)$ of degree 1 or higher has a root z where $p(z) = 0$.

Notice that a first-degree polynomial like $3x - 6$ can be factored as $3(x - 2)$, where 2 is the only root of $3x - 6$. In general, if $a \neq 0$, the polynomial $ax - b$ can be factored as $a(x - (b/a))$ where b/a is the root of $ax - b$.

Similarly, for every second-degree polynomial $ax^2 + bx + c$, we can factor it as $a(x - z_1)(x - z_2)$ where z_1 and z_2 are (possibly complex, possibly the same) roots of the polynomial. As a consequence of the fundamental theorem of algebra, this pattern extends to polynomials of any degree.

Corollary: Every polynomial of degree $n \geq 1$ can be factored into n parts. More specifically, if $p(x)$ is an nth-degree polynomial with leading term $a \neq 0$, then there exist n numbers (possibly complex, possibly the same) $z_1, z_2, \ldots, z_n$ such that $p(x) = a(x - z_1)(x - z_2) \cdots (x - z_n)$. The numbers z_i are the roots of the polynomial where $p(z_i) = 0$.

This corollary means that every polynomial of degree $n \geq 1$ has at least one, and at most n, distinct roots. For example, the polynomial $x^4 - 16$ has degree 4 and can be factored as

$$x^4 - 16 = (x^2 - 4)(x^2 + 4) = (x - 2)(x + 2)(x - 2i)(x + 2i)$$

and has four distinct roots $2, -2, 2i, -2i$. The polynomial $3x^3 + 9x^2 - 12$ has degree 3, but since it factors as

$$3x^3 + 9x^2 - 12 = 3(x^2 + 4x + 4)(x - 1) = 3(x + 2)^2(x - 1)$$

it has only two distinct roots, -2 and 1.

The Geometry of Complex Numbers

The complex numbers can be visualized by drawing the *complex plane*. It looks just like the (x, y) plane from algebra, but the y-axis has been replaced with the *imaginary axis* with numbers like $0, \pm i, \pm 2i$, and so on. We have plotted some complex numbers in the figure below.

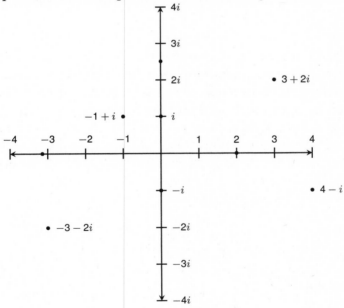

Some points in the complex plane

We have seen how easy it is to add, subtract, and multiply complex numbers numerically. But we can also perform these operations geometrically, just by looking at their points on the complex plane.

For example, consider the addition problem

$$(3 + 2i) + (-1 + i) = 2 + 3i$$

In the figure below, notice that the points 0, $3 + 2i$, $2 + 3i$, and $-1 + i$ form the vertices of a parallelogram.

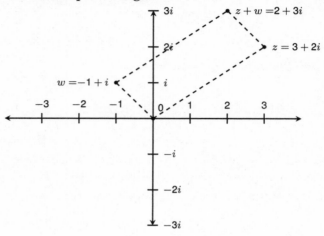

In general, we can add complex numbers z and w geometrically, just by drawing a parallelogram, as in the previous example. To do the subtraction problem $z - w$, we plot the point $-w$ (which is located symmetrically opposite w) and add the points z and $-w$, as illustrated below.

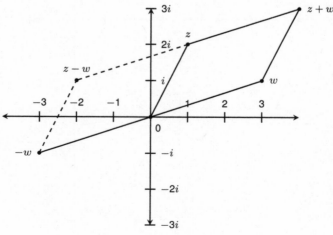

Complex numbers can be added and subtracted by drawing parallelograms

In order to multiply and divide complex numbers geometrically, we first need to measure their sizes. We define the *length* (or *magnitude*) of a complex number z, denoted $|z|$, to be the length of the line segment from the origin 0 to the point z. Specifically, if $z = a + bi$, then by the Pythagorean theorem, z has length

$$|z| = \sqrt{a^2 + b^2}$$

For example, as illustrated below, the point $3 + 2i$ has length $\sqrt{3^2 + 2^2} = \sqrt{13}$. Note that the angle θ corresponding to $3 + 2i$ would satisfy $\tan \theta = 2/3$. Thus $\theta = \tan^{-1} 2/3 \approx 33.7°$ or about 0.588 radians.

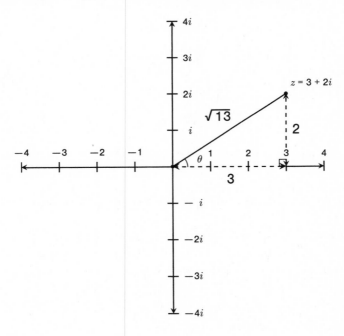

The complex number $z = 3 + 2i$ has length $|z| = \sqrt{13}$, and an angle θ where $\tan \theta = 2/3$

If you plot the points that have length 1, you get the *unit circle* on the complex plane, shown on the next page. What is the complex number on the circle associated with angle θ? If this were the x-y Cartesian plane, then from Chapter 9 we know that it would be the point $(\cos \theta, \sin \theta)$. So in the complex plane, this would be $\cos \theta + i \sin \theta$. Likewise, any complex number with length R is of the form

$$z = R(\cos \theta + i \sin \theta)$$

We call this the *polar form* of the complex number. Maybe I shouldn't tell you this now, but at the end of this chapter, we will learn that this is also equal to $Re^{i\theta}$. (Would that be a "spEuler alert"?)

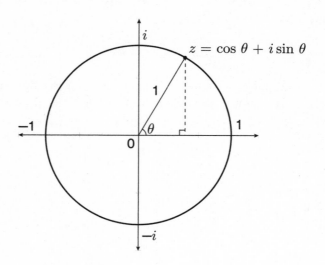

The unit circle in the complex plane

Remarkably, when complex numbers are multiplied, their lengths multiply as well.

Theorem: For complex numbers z_1 and z_2, $|z_1 z_2| = |z_1||z_2|$. In other words, *the length of the product is the product of the lengths.*

✗ Aside

Proof: Let $z_1 = a + bi$ and $z_2 = c + di$. Then $|z_1| = \sqrt{a^2 + b^2}$ and $|z_2| = \sqrt{c^2 + d^2}$. Thus,

$$
\begin{aligned}
|z_1 z_2| &= |(a + bi)(c + di)| = |(ac - bd) + (ad + bc)i| \\
&= \sqrt{(ac - bd)^2 + (ad + bc)^2} \\
&= \sqrt{(ac)^2 + (bd)^2 - 2abcd + (ad)^2 + (bc)^2 + 2abcd} \\
&= \sqrt{(ac)^2 + (bd)^2 + (ad)^2 + (bc)^2} \\
&= \sqrt{(a^2 + b^2)(c^2 + d^2)} \\
&= \sqrt{a^2 + b^2} \sqrt{c^2 + d^2} \\
&= |z_1||z_2|
\end{aligned}
$$

□

For example,

$$|(3 + 2i)(1 - 3i)| = |9 - 7i| = \sqrt{9^2 + (-7)^2} = \sqrt{130}$$
$$= \sqrt{13}\sqrt{10} = |3 + 2i|\,|1 - 3i|$$

What about the angle of the product? The notation $\arg z$ is often used to denote the angle that the complex number z makes with the positive x-axis. For instance, we saw that $\arg(3 + 2i) = 0.588$ radians. Likewise, since $1 - 3i$ is in quadrant IV and its angle satisfies $\tan\theta = -3$, we have $\arg(1 - 3i) = \tan^{-1}(-3) = -71.56° = -1.249$ radians.

Note that $(3 + 2i)(1 - 3i) = (9 - 7i)$ has angle $\tan^{-1}(-7/9) = -37.87° = -0.661$ radian, which just happens to be $0.588 + (-1.249)$. According to the next theorem, this is not a coincidence!

Theorem: For complex numbers z_1 and z_2, $\arg(z_1 z_2) = \arg(z_1) + \arg(z_2)$. In other words, *the angle of the product is the sum of the angles.*

The proof, presented in the following box, relies on trigonometric identities from the previous chapter.

✂ Aside

Proof: Suppose that z_1 and z_2 are complex numbers with respective lengths R_1 and R_2 and respective angles θ_1, and θ_2. Then, writing z_1 and z_2 in polar form, we have

$$z_1 = R_1(\cos\theta_1 + i\sin\theta_1) \qquad z_2 = R_2(\cos\theta_2 + i\sin\theta_2)$$

Therefore,

$$z_1 z_2 = R_1(\cos\theta_1 + i\sin\theta_1)R_2(\cos\theta_2 + i\sin\theta_2)$$
$$= R_1 R_2[\cos\theta_1\cos\theta_2 - \sin\theta_1\sin\theta_2 + i(\sin\theta_1\cos\theta_2 + \sin\theta_2\cos\theta_1)]$$
$$= R_1 R_2[\cos(\theta_1 + \theta_2) + i(\sin(\theta_1 + \theta_2))]$$

where we exploited the identities for $\cos(A + B)$ and $\sin(A + B)$, which were derived last chapter. Consequently, $z_1 z_2$ has length $R_1 R_2$ (which we knew) and angle $\theta_1 + \theta_2$, as was to be shown. □

To summarize, when multiplying complex numbers, you simply *multiply their lengths and add their angles.* For example, when multiplying a number by i, the length stays the same, but the angle increases by 90°. Notice that when we multiply real numbers together, the positive numbers have angles of 0° (or equivalently, 360°) and the negative numbers have angles of 180°. When you add angles of 180° together, you get an angle of 360°, which is another way of saying that the product of two negative numbers is a positive number. The imaginary numbers have angles of 90° and −90° (or 270°). Thus, when you multiply an imaginary number by itself, the angle must be 180° (since 90° + 90° = 180° or −90° + −90° = −180° is the same as 180°), which is a negative number. Finally, note that if z has angle θ, then $1/z$ must have angle $-\theta$. (Why? Since $z \cdot 1/z = 1$, the angles for z and $1/z$ must sum to 0°.) Therefore, when *dividing* complex numbers, you *divide* their lengths and *subtract* their angles. That is, z_1/z_2 has length R_1/R_2 and angle $\theta_1 - \theta_2$.

We're sorry. You have reached an imaginary number. If you need a real number, then please rotate your phone by 90 degrees and try again!

The Magic of e

If you have a scientific calculator, please try the following experiment.

1. Enter a memorable seven-digit number on your calculator (perhaps a phone number or identification number, or maybe your favorite one-digit number repeated seven times).

2. Take the reciprocal of that number (by pressing the $1/x$ button on your calculator).

3. Add 1 to your answer.

4. Now raise this number to the power of your original seven-digit number (by pressing the x^y button, followed by your seven-digit number, followed by the equals sign).

Does your answer begin 2.718? In fact, it wouldn't surprise me if your answer began with several digits of the irrational number

$$e = 2.718281828459045\ldots$$

So what is this mysterious number e, and why is it so important? In the magic trick you just performed, you calculated

$$(1 + 1/n)^n$$

for some large number n. Now, what would you expect to happen to this number as n gets larger and larger? On the "one" hand, as n gets larger, the number $(1 + 1/n)$ gets closer and closer to the number 1, and when we raise 1 to any power, we still get 1. Thus it would be reasonable to expect that for very large values of n, $(1 + 1/n)^n$ would be approximately 1. For example, $(1.001)^{100} \approx 1.105$.

On the other hand, even when n is large, $(1 + 1/n)$ is still slightly bigger than 1. And if you raise any fixed number that is bigger than 1 to larger and larger powers, then that quantity gets arbitrarily large. For example, $(1.001)^{10,000}$ is over 20,000.

The problem is that the base $(1 + 1/n)$ is getting small while the exponent n is getting large *simultaneously*. And in this tug-of-war between 1 and infinity, the answer gets closer and closer to $e = 2.71828\ldots$. For example, $(1.001)^{1000} \approx 2.717$. Let's look at the function $(1 + 1/n)^n$ for large values of n, as shown in the following table.

n	$(1 + 1/n)^n$
10	$(1.1)^{10} = 2.5937424\ldots$
100	$(1.01)^{100} = 2.7048138\ldots$
1000	$(1.001)^{1000} = 2.7169239\ldots$
10,000	$(1.0001)^{10,000} = 2.7181459\ldots$
100,000	$(1.00001)^{100,000} = 2.7182682\ldots$
1,000,000	$(1.000001)^{1,000,000} = 2.7182805\ldots$
10,000,000	$(1.0000001)^{10,000,000} = 2.7182817\ldots$

We define e to be the number that $(1 + 1/n)^n$ is getting closer and closer to as n gets larger and larger. Mathematicians call this the *limit* of $(1 + 1/n)^n$ as n goes to infinity, denoted

$$e = \lim_{n \to \infty} (1 + 1/n)^n$$

If we replace the fraction $1/n$ with x/n where x is any real number, then as n/x gets larger and larger, the number $(1 + x/n)^{n/x}$ gets closer and closer to e. Raising both sides to the power x (and recalling that $(a^b)^c = a^{bc}$) gives us what's called the *exponential formula*:

$$\lim_{n \to \infty} (1 + x/n)^n = e^x$$

The exponential formula has many *interesting* applications. Suppose you put 10,000 dollars in a savings account that earns an interest rate of 0.06 (that is, 6 percent per year). If the interest is applied annually, then at the end of one year, you would have $10,000(1.06) = 10,600$ dollars. After two years, you would earn 6 percent on this new amount and have $10,000(1.06)^2 = 11,236$ dollars. In three years, you would have $10,000(1.06)^3 = 11,910.16$ dollars. After t years, you would have

$$10,000(1.06)^t$$

dollars. More generally, if we replace the interest rate of 0.06 with an interest rate r, and if you begin with a principal of P dollars, then at the end of t years, the number of dollars you would have would be

$$P(1 + r)^t$$

Now suppose our 6 percent interest is compounded semiannually: so you earn 3 percent every six months. Then after a year you would have $10,000(1.03)^2 = 10,609$ dollars, which is a little more than the 10,600 dollars for annual compounding. If the money is compounded quarterly, then you earn 1.5 percent four times a year, yielding $10,000(1.015)^4 = 10,613.63$ dollars. More generally, if our money is compounded n times per year, then after one year, you would have

$$10,000 \left(1 + \frac{0.06}{n}\right)^n$$

dollars. Letting n get very large is called *continuous* compounding. By the exponential formula, after one year you would have

$$10,000 \lim_{n \to \infty} \left(1 + \frac{0.06}{n}\right)^n = 10,000 e^{0.06} = 10,618.36$$

dollars, as shown in the next table.

Principal	Interest	Compounded	Amount after one year
$10,000	6%	Annually	$10,000(1.06) = $10,600.00
$10,000	6%	Semiannually	$10,000(1.03)^2 = $10,609.00
$10,000	6%	Quarterly	$10,000(1.015)^4 = $10,613.83
$10,000	6%	Monthly	$10,000(1.005)^{12} = $10,616.77
$10,000	6%	n installments	$10,000(1 + \frac{0.06}{n})^n
$10,000	6%	Continuously	$10,000e^{0.06} = $10,618.36

More generally, if you start with an initial principal of P dollars and if your money is compounded continuously at interest rate r, then after t years, you will have A dollars given by this pretty (or perhaps I should say "pertly") formula:

$$A = Pe^{rt}$$

As seen in the graph below, the function $y = e^x$ grows very quickly. Alongside it, we also present the graphs of e^{2x} and $e^{0.06x}$. We say that these functions grow *exponentially*. The graph $y = e^{-x}$ goes to 0 very quickly and exhibits *exponential decay*.

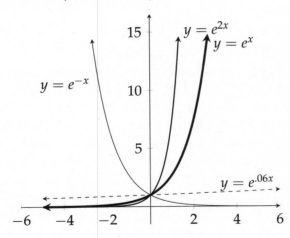

Some exponential functions

How about the graph of 5^x? Since $e < 5 < e^2$, then 5^x must lie in between the functions e^x and e^{2x}. More specifically, it turns out that $e^{1.609\ldots} = 5$ and therefore $5^x \approx e^{1.609x}$. In general, any function a^x can be expressed as an exponential function e^{kx}, once we find an exponent k for which $a = e^k$. How do we find k? Using *logarithms*.

In the same way that the square root is the inverse of the squaring function (since the functions undo each other), the logarithm is the inverse of the exponential function. The most commonly used logarithm is the base 10 logarithm, denoted $\log x$. We say that

$$y = \log x \quad \text{if} \quad 10^y = x$$

Or equivalently,

$$10^{\log x} = x$$

For example, since $10^2 = 100$, we have $\log 100 = 2$. Here is a useful table of logarithms.

Logarithm	Explanation
$\log 1 = 0$	Since $10^0 = 1$
$\log 10 = 1$	Since $10^1 = 10$
$\log 100 = 2$	Since $10^2 = 100$
$\log 1000 = 3$	Since $10^3 = 1000$
$\log(1/10) = -1$	Since $10^{-1} = 1/10$
$\log .01 = -2$	Since $10^{-2} = .01$
$\log \sqrt{10} = 1/2$	Since $10^{1/2} = \sqrt{10}$
$\log 10^x = x$	Since $10^x = 10^x$
$\log 0$ is undefined	Since no y has $10^y = 0$

One of the reasons that logarithms are so useful is that they transform large numbers into much smaller numbers that our brains are better able to comprehend. For example, the Richter scale uses logarithms, which allow us to measure the size of an earthquake on a scale of 1 to 10. Logarithms are also used for measuring the intensity of sound (using decibels), the acidity of a chemical solution (pH), and even the popularity of a webpage through Google's PageRank algorithm.

What is $\log 512$? Any scientific calculator (or even most search engines) will tell you $\log 512 = 2.709\ldots$. This seems reasonable: since 512 lies between 10^2 and 10^3, its logarithm must be between 2 and 3. Logarithms were invented as a tool for converting multiplication problems into easier addition problems. This was based on the following useful theorem.

Theorem: For any positive numbers x and y,

$$\log xy = \log x + \log y$$

In other words, the log of the product is the sum of the logs.

Proof: This comes immediately from the law of exponents, since

$$10^{\log x + \log y} = 10^{\log x} 10^{\log y} = xy = 10^{\log xy}$$

Thus, raising 10 to the $\log x + \log y$ power gives us xy, as desired.   □

Another useful property is the *exponent rule*.

Theorem: For any positive number x and any integer n,

$$\log x^n = n \log x$$

Proof: By the law of exponents, $a^{bc} = (a^b)^c$. Therefore,

$$10^{n \log x} = (10^{\log x})^n = x^n$$

Hence the logarithm of x^n equals $n \log x$.   □

There is nothing particularly special about the base 10 logarithm, although it is widely used in chemistry and physical sciences like geology. But in computer science and discrete mathematics, the base 2 logarithm is more popular. For any $b > 0$, the base b logarithm $\log_b$ is defined by the rule

$$y = \log_b x \text{ if } b^y = x$$

For example, $\log_2 32 = 5$ since $2^5 = 32$. All of the previous logarithm properties hold for any base b. For example,

$$b^{\log_b x} = x \qquad \log_b xy = \log_b x + \log_b y \qquad \log_b x^n = n \log_b x$$

However, in most areas of mathematics, physics, and engineering, the most useful logarithm is with base $b = e$. This is called the *natural logarithm* and is denoted by $\ln x$. That is,

$$y = \ln x \text{ if } e^y = x$$

Or equivalently, for any real number x,

$$\ln e^x = x$$

For example, your calculator can determine $\ln 5 = 1.609\ldots$, which is how we determined earlier that $e^{1.609} \approx 5$. We will have more to say about the natural logarithm function in Chapter 11.

> ✗ **Aside**
>
> All scientific calculators compute natural logarithms and base 10 logarithms, but most do not explicitly calculate logarithms in other bases. But this turns out to not be a problem, since there is a simple way to convert logarithms from one base to another. Essentially, if you know one logarithm, you know them all. Specifically, using just the base 10 logarithm, we can determine the base b logarithm with the following rule.
>
> **Theorem:** For any positive numbers b and x,
>
> $$\log_b x = \frac{\log x}{\log b}$$
>
> **Proof:** Let $y = \log_b x$. Then $b^y = x$. Taking the log of both sides, $\log b^y = \log x$. And by the exponent rule, this says $y \log b = \log x$. Therefore $y = (\log x)/(\log b)$, as desired.   □
> For example, for any $x > 0$,
>
> $$\ln x = (\log x)/(\log e) = (\log x)/(0.434\ldots) \approx 2.30 \log x$$
>
> $$\log_2 x = (\log x)/(\log 2) = (\log x)/(0.301\ldots) \approx 3.32 \log x$$

More Appearances of e

Just like the number π, the number e is pervasive in mathematics, showing up in places where you wouldn't expect it. For example, the classic bell curve, which we saw in Chapter 8, has the formula

$$y = \frac{e^{-x^2/2}}{\sqrt{2\pi}}$$

Its graph (shown opposite) is probably the most important graph in the subject of statistics.

In Chapter 8 we also saw e appear in Stirling's approximation for $n!$:

$$n! \approx \left(\frac{n}{e}\right)^n \sqrt{2\pi n}$$

As we'll see in Chapter 11, e is fundamentally connected to the factorial function. We will show that e^x has the infinite series

$$e^x = 1 + \frac{x}{1!} + \frac{x^2}{2!} + \frac{x^3}{3!} + \frac{x^4}{4!} + \cdots$$

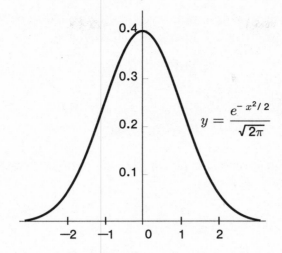

The bell curve has formula $e^{-x^2/2}/\sqrt{2\pi}$

In particular, when $x = 1$, this formula says

$$e = 1 + 1 + \frac{1}{2!} + \frac{1}{3!} + \frac{1}{4!} + \cdots$$

which is a very quick way to determine the digits of e.

By the way, the digits of e begin with the repetitive pattern

$$e = 2.718281828\ldots$$

or as my high school teacher would say, "2.7 Andrew Jackson, Andrew Jackson," since 1828 was the year that the seventh US president was elected. (Although for me the mnemonic worked the other way. I remember the year of Jackson's election from the digits of e.) You might be tempted to believe that e is a rational number, which it would be if the digit sequence 1828 repeated forever, but that is not the case. The next six digits of e are ...459045..., which I remember as the angles of an isosceles right triangle.

The number e also shows up in many probability problems where you wouldn't expect to see it. For example, let's suppose that every week you purchase a raffle ticket where your chance of winning a prize is 1 in 100. If you buy a ticket for 100 consecutive weeks, what are your chances of winning a prize at least once? Each week, your probability of winning is $1/100 = 0.01$ and your chance of losing is $99/100 = 0.99$.

Since your chance of winning in any given week is independent of previous weeks, your chance of losing all 100 weeks is

$$(0.99)^{100} \approx 0.3660$$

which is very close to

$$1/e \approx 0.3678794\ldots$$

This is not a coincidence. Recalling the exponential formula when we first introduced e^x, we have

$$\lim_{n \to \infty} \left(1 + \frac{x}{n}\right)^n = e^x$$

Now if we let $x = -1$, then for any large number n, we have

$$\left(1 - \frac{1}{n}\right)^n \approx e^{-1} = 1/e$$

When $n = 100$, this says that $(0.99)^{100} \approx 1/e$, as promised. Thus your chance of winning is about $1 - (1/e) \approx 64$ percent.

One of my favorite probability problems goes by the name of the *matching problem* (or the hat-check problem or the derangement problem). Suppose that n homework assignments are being returned to a class, but the teacher is lazy and gives each student a random assignment (which may be that student's or may belong to any of the other students in the class). What is the probability that no student receives her own homework back? Equivalently, if the numbers 1 through n are randomly mixed, what is the probability that no number is in its natural position? For example, when $n = 3$, the numbers $1, 2, 3$ can be arranged $3! = 6$ ways, and there are 2 *derangements* where no number is in its natural position, namely 231 and 312. Thus when $n = 3$, the probability of a derangement is $2/6 = 1/3$.

With n homework assignments being returned, there are $n!$ possible ways to return them. If we let D_n denote the number of derangements, then the probability that nobody gets her own homework back is $p_n = D_n/n!$. For example, when $n = 4$, there are 9 derangements:

2143   2341   2413   3142   3412   3421   4123   4312   4321

Thus $p_4 = D_4/4! = 9/24 = 0.375$, as listed in the following table.

n	D_n	$p_n = D_n/n!$
1	0	0
2	1	1/2 = 0.50000
3	2	2/6 = 0.33333
4	9	9/24 = 0.37500
5	44	44/120 = 0.36667
6	265	265/720 = 0.36806
7	1856	1865/5040 = 0.36825
8	14,887	14,887/40,320 = 0.36823

As n gets larger and larger, p_n will get closer and closer to $1/e$. The implications are astounding. This says that the likelihood of nobody receiving her own homework back is virtually the same, whether the class has 10 students or 100 students or one million students! The chance is really, really close to $1/e$.

Where does $1/e$ come from? As a first approximation, with n students, each student has a $1/n$ chance of receiving her homework back and therefore a $1 - (1/n)$ chance of getting someone else's assignment. Thus the probability that all n students get someone else's assignment is

$$p_n \approx \left(1 - \frac{1}{n}\right)^n \approx 1/e$$

The probability is approximate because, unlike the raffle problem, we do not quite have independent events. If student 1 gets her own homework, that slightly increases the probability that student 2 gets his. (The probability would be $1/(n-1)$ instead of $1/n$.) Likewise, if student 1 does not get her homework back, then student 2's chances go down ever so slightly. But since the probabilities don't change too much, the approximation is very good.

The exact probability for p_n uses the infinite series for e^x,

$$e^x = 1 + x + \frac{x^2}{2!} + \frac{x^3}{3!} + \frac{x^4}{4!} + \cdots$$

When we substitute $x = -1$ into this equation, we get

$$1 - 1 + \frac{1}{2!} - \frac{1}{3!} + \frac{1}{4!} - \cdots = e^{-1} = 1/e$$

It can be shown that for n students, the probability that nobody receives their own homework is exactly

$$p_n = 1 - \frac{1}{1!} + \frac{1}{2!} - \frac{1}{3!} + \frac{1}{4!} - \cdots + (-1)^n \frac{1}{n!}$$

For example, with $n = 4$ students, $p_n = 1 - 1 + 1/2 - 1/6 + 1/24 = 9/24$, as previously shown. The convergence to $1/e$ is extremely fast. The distance between p_n and $1/e$ is less than $1/(n+1)!$. Thus p_4 is within $1/5! = 0.0083$ of $1/e$; p_{10} agrees with $1/e$ to seven decimal places; p_{100} agrees with $1/e$ to over 150 decimal places!

✕ Aside

Theorem: The number e is irrational.

Proof: Suppose, to the contrary, that e is rational. Then $e = m/n$ for some positive integers m and n. Now let's use the number n to split the infinite series for e into two parts, so that $e = L + R$, where

$$L = 1 + 1 + \frac{1}{2!} + \frac{1}{3!} + \frac{1}{4!} + \cdots + \frac{1}{(n-1)!} + \frac{1}{n!}$$

$$R = \frac{1}{(n+1)!} + \frac{1}{(n+2)!} + \frac{1}{(n+3)!} + \cdots$$

Notice that $n!e = en(n-1)! = m(n-1)!$ must be an integer (since m and $(n-1)!$ are integers) and $n!L$ is an integer too (since $n!/k!$ is an integer for all $k \leq n$). Thus $n!R = n!e - n!L$ is the difference of two integers, so it must be an integer itself. But this is impossible, since $n \geq 1$ implies that

$$
\begin{aligned}
n!R \;&=\; \frac{1}{n+1} + \frac{1}{(n+1)(n+2)} + \frac{1}{(n+1)(n+2)(n+3)} + \cdots \\
&\leq\; \frac{1}{2} + \frac{1}{2 \cdot 3} + \frac{1}{2 \cdot 3 \cdot 4} + \cdots \\
&=\; \frac{1}{2!} + \frac{1}{3!} + \frac{1}{4!} + \cdots = 0.71828... \\
&<\; 1
\end{aligned}
$$

So $n!R$ can't be an integer because there are no positive integers less than 1. Hence the assumption that $e = m/n$ leads to a contradiction, and therefore e is irrational. □

Euler's Equation

The number e was explored and popularized by the great mathematician Leonhard Euler, and he was the first to assign this fundamental number its current name. Most historians of mathematics disagree with the suggestion that he chose the letter e because it was the first letter of his last name. But many people still refer to e as Euler's number.

We have already introduced the infinite series for the functions e^x, $\cos x$, and $\sin x$, and we will explain where they come from in the next chapter. But let's put them all in one place here.

$$e^x = 1 + x + \frac{x^2}{2!} + \frac{x^3}{3!} + \frac{x^4}{4!} + \cdots$$

$$\cos x = 1 - \frac{x^2}{2!} + \frac{x^4}{4!} - \frac{x^6}{6!} + \cdots$$

$$\sin x = x - \frac{x^3}{3!} + \frac{x^5}{5!} - \frac{x^7}{7!} + \cdots$$

These formulas are valid for all real numbers x, but Euler had the audacity to imagine what they would say if we let x be an imaginary number. What would it mean for a number to be raised to an imaginary power? The result is Euler's beautiful theorem.

Theorem (Euler's theorem): For any angle θ (measured in radians),

$$e^{i\theta} = \cos \theta + i \sin \theta$$

Proof: We prove this theorem by seeing what happens when we substitute $x = i\theta$ in the series for e^x.

$$e^{i\theta} = 1 + i\theta + \frac{(i\theta)^2}{2!} + \frac{(i\theta)^3}{3!} + \frac{(i\theta)^4}{4!} + \frac{(i\theta)^5}{5!} + \frac{(i\theta)^6}{6!} + \frac{(i\theta)^7}{7!} + \cdots$$

Note what happens to the power of i as it is raised to various powers: $i^0 = 1, i^1 = i, i^2 = -1, i^3 = -i$ (since $i^3 = i^2 i = -i$), and then the pattern repeats: $i^4 = 1, i^5 = i, i^6 = -1, i^7 = -i, i^8 = 1$, and so on. In particular, notice that the powers of i alternate between real and imaginary, and we can factor the number i out of every second term, as in the algebra that follows.

$$
\begin{aligned}
e^{i\theta} &= 1 + i\theta - \frac{\theta^2}{2!} - i\frac{\theta^3}{3!} + \frac{\theta^4}{4!} + i\frac{\theta^5}{5!} - \frac{\theta^6}{6!} - i\frac{\theta^7}{7!} + \frac{\theta^8}{8!} + \cdots \\
&= \left(1 - \frac{\theta^2}{2!} + \frac{\theta^4}{4!} - \frac{\theta^6}{6!} + \cdots\right) + i\left(\theta - \frac{\theta^3}{3!} + \frac{\theta^5}{5!} - \frac{\theta^7}{7!} + \cdots\right) \\
&= \cos \theta + i \sin \theta \qquad \qquad \qquad \qquad \qquad \qquad \quad ☺
\end{aligned}
$$

This gives us the proof of "God's equation," introduced at the beginning of the chapter. Letting $\theta = \pi$ radians (or $180°$), we have

$$e^{i\pi} = \cos\pi + i\sin\pi = -1 + i(0) = -1$$

But Euler's theorem says much more than this. We have seen the expression $\cos\theta + i\sin\theta$ before. It is the point on the unit circle of the complex plane that has an angle of θ relative to the positive x-axis. Euler's theorem says that you can express that point in a simple way, as shown in the figure.

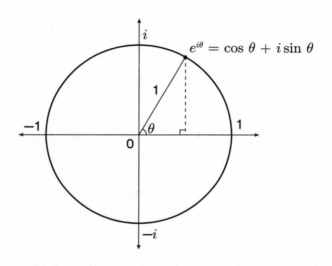

By Euler's theorem, the points on the unit circle are all of the form $e^{i\theta}$

But wait, there's more! Every point on the complex plane is just a scaled version of a point on the unit circle. Specifically, if the complex number z has a length of R and an angle of θ, then that point is just R times the corresponding point on the unit circle. In other words,

$$z = Re^{i\theta}$$

Thus if we have any two points in the complex plane, say $z_1 = R_1 e^{i\theta_1}$ and $z_2 = R_2 e^{i\theta_2}$, then the law of exponents (with complex numbers) tells us that

$$z_1 z_2 = R_1 e^{i\theta_1} R_2 e^{i\theta_2} = R_1 R_2 e^{i(\theta_1 + \theta_2)}$$

which is the complex number with length $R_1 R_2$ and angle $\theta_1 + \theta_2$. So once again we can conclude that to multiply complex numbers, you simply multiply their lengths and add their angles. When we proved this fact earlier in the chapter, we relied on about a page's worth of algebra and trigonometric identities. But with Euler's theorem, we arrive at this conclusion in just one line, all thanks to the number e!

Let's end with a poem to celebrate this remarkable number, with apologies to Joyce Kilmer.

> I think that I shall never see
> A number lovelier than e.
> Whose digits are too great to state
> They're 2.71828 . . .
> And e has such amazing features.
> It's loved by all (but mostly teachers).
> With all of e's great properties,
> Most integrals are done with . . . *ease.*
> Theorems are proved by fools like me.
> But only Euler could make an e.

$$y = x^{11} \implies y' = 11x^{10}$$

The Magic of Calculus

Going off on Tangents

Mathematics is the language of science, and the mathematics used to express most laws of nature is calculus. Calculus is the mathematics of how things grow and change and move. In this chapter, we will learn how to determine the rate at which functions change and how to approximate complicated functions with simpler functions like polynomials. Calculus is also a powerful tool for *optimization* and can be useful for determining how to choose your numbers in such a way as to maximize a quantity (like profit or volume) or minimize a quantity (like cost or distance traveled).

For example, suppose you have a square piece of cardboard, 12 inches per side, as shown on the next page. Suppose you cut x-by-x squares out of the corners and then fold the resulting tabs up to create a tray. What is the maximum possible volume of the resulting tray?

Let's start by computing the volume as a function of x. The base of the tray would have area $(12 - 2x)(12 - 2x)$ and the height of the tray would be x, so the volume of the tray would be

$$V = (12 - 2x)^2 x$$

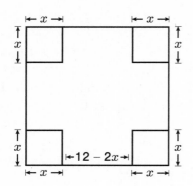 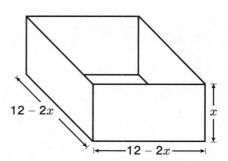

What value of x maximizes the volume of the box?

cubic inches. Our goal is to choose the value of x to make this volume as large as possible. We can't choose x to be too big or too small. For instance, if $x = 0$ or $x = 6$, then the box has a volume of 0. The optimal value of x lies somewhere in between.

Below is a graph of the function $y = (12 - 2x)^2 x$ as x ranges from 0 to 6. When $x = 1$, we compute that the volume is $y = 100$. When $x = 2$, $y = 128$. When $x = 3$, $y = 108$. The value of $x = 2$ looks promising, but perhaps there is a real number that does even better somewhere between 1 and 3?

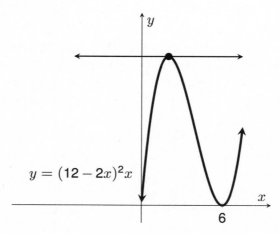

The point where $y = (12 - 2x)^2 x$ is maximized has a horizontal tangent line

Just to the left of the maximum, the function is going uphill, with a positive slope, and just to the right, it is going downhill, with a negative slope. So, at the maximum point, the function is neither increasing nor decreasing: it is switching between the two. To put it more mathematically, at this optimal point, there is a horizontal tangent line (with a slope of 0). In this chapter, we will use calculus to find that point between 0 and 6 where the tangent line is horizontal.

And speaking of tangents, we will be going off on many tangents throughout this chapter. For instance, the problem we just considered was to find the optimal way to cut corners, and indeed we will be cutting lots of corners in this chapter. Calculus is a vast subject, with typical textbooks containing more than a thousand pages. In just a couple dozen pages, we will only have time to cover the highlights. In this book, we will not cover the topic of *integral calculus*, which computes areas and volumes of complicated objects; we will focus only on *differential calculus*, which measures how functions grow and change.

The simplest functions to analyze are straight lines. In Chapter 2, we noted that the line $y = mx + b$ has a slope of m. Thus if x increases by 1, then y increases by m. For example, the line $y = 2x + 3$ has a slope of 2. If we increase the value of x by 1 (say from $x = 10$ to $x = 11$), then y will increase by 2 (here, from 23 to 25).

We have drawn the graphs of various lines in the figure below. Here, $y = -x$ has slope -1, and the horizontal line $y = 5$ has slope 0.

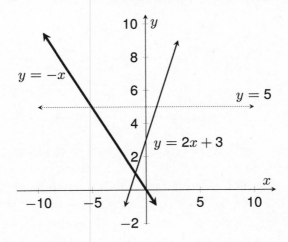

Graphs of lines

Given any two points, we can draw a line through them and determine the slope of that line without needing the line's formula. The *slope* of the line that goes through the point (x_1, y_1) and (x_2, y_2) is given by the "rise over run" formula:

$$m = \frac{y_2 - y_1}{x_2 - x_1}$$

For example, take any two points on the line $y = 2x + 3$, say the points $(0, 3)$ and $(4, 11)$. Then the slope of the line connecting these points is $m = \frac{y_2 - y_1}{x_2 - x_1} = (11 - 3)/(4 - 0) = 8/4 = 2$, which is exactly the slope we see in the original equation of that line.

Now consider the function $y = x^2 + 1$, as shown in the graph below. This graph is not a straight line and we can see that the slope is always changing. Let's try to determine the slope of the tangent line at the point $(1, 2)$.

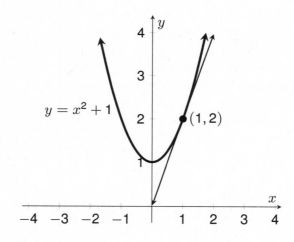

For $y = x^2 + 1$, find the slope of the tangent line at the point $(1, 2)$

The bad news is that it takes two points to determine a slope, and we only have the one point $(1, 2)$. Thus we first approximate the slope of the tangent line by looking at a line that goes through two points on the curve (called a *secant* line), as shown on the right. If $x = 1.5$, then $y = (1.5)^2 + 1 = 3.25$. So let's look at the slope of the line from $(1, 2)$ to $(1.5, 3.25)$. According to our slope formula, the slope of this secant line is

$$m = \frac{y_2 - y_1}{x_2 - x_1} = \frac{3.25 - 2}{1.5 - 1} = \frac{1.25}{0.5} = 2.5$$

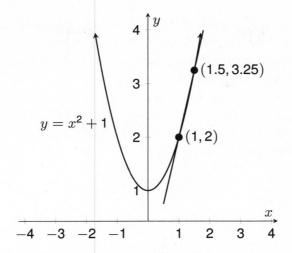

Approximating the tangent line with a secant line

For a better approximation, we move the second point closer to $(1, 2)$. For instance, if $x = 1.1$, then $y = (1.1)^2 + 1 = 2.21$, and the resulting secant slope is $m = (2.21 - 2)/(1.1 - 1) = 2.1$. As shown in the table below, as we move the second point closer and closer to $(1, 2)$, the secant slope seems to get closer and closer to 2.

(x_1, y_1)	x_2	$y_2 = x_2^2 + 1$	$\frac{y_2 - y_1}{x_2 - x_1}$		Slope
$(1, 2)$	1.5	3.25	$\frac{3.25 - 2}{1.5 - 1} = \frac{1.25}{0.5}$	$=$	2.5
$(1, 2)$	1.1	2.21	$\frac{2.21 - 2}{1.1 - 1} = \frac{0.21}{0.1}$	$=$	2.1
$(1, 2)$	1.01	2.0201	$\frac{2.0201 - 2}{1.01 - 1} = \frac{0.0201}{0.01}$	$=$	2.01
$(1, 2)$	1.001	2.002001	$\frac{2.002001 - 2}{1.001 - 1} = \frac{0.002001}{0.001}$	$=$	2.001
$(1, 2)$	$1 + h$	$2 + 2h + h^2$	$\frac{(2 + 2h + h^2) - 2}{(1 + h) - 1} = \frac{2h + h^2}{h}$	$=$	$2 + h$

Look what happens when $x = 1 + h$ where $h \neq 0$, but could be just a *hair's* length away from $x = 1$. Then $y = (1 + h)^2 + 1 = 2 + 2h + h^2$. The slope of the secant line would then be

$$\frac{y_2 - y_1}{x_2 - x_1} = \frac{(2 + 2h + h^2) - 2}{(1 + h) - 1} = \frac{2h + h^2}{h} = 2 + h$$

Now as h gets closer and closer to 0, the secant slope gets closer and closer to 2. Formally, we say

$$\lim_{h \to 0} (2 + h) = 2$$

This notation means that the *limit* of $2 + h$ as h goes to zero is 2. Intuitively, as h gets closer and closer to 0, $2 + h$ gets closer and closer to 2. So we have found that for the graph $y = x^2 + 1$ at the point $(1, 2)$, the slope of the tangent line is 2.

The general situation looks like this. For the function $y = f(x)$, we want to find the slope of the tangent line at the point $(x, f(x))$. As pictured below, the slope of the secant line through the point $(x, f(x))$ and the neighboring point $(x + h, f(x + h))$ is

$$\frac{y_2 - y_1}{x_2 - x_1} = \frac{f(x + h) - f(x)}{(x + h) - x} = \frac{f(x + h) - f(x)}{h}$$

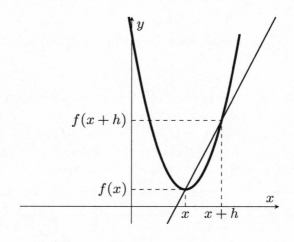

The slope of the secant line through $(x, f(x))$ and $(x + h, f(x + h))$ is $\frac{f(x+h) - f(x)}{h}$

We use the notation $f'(x)$ to denote the slope of the tangent line at the point $(x, f(x))$, so

$$f'(x) = \lim_{h \to 0} \frac{f(x + h) - f(x)}{h}$$

This is a complicated definition, so let's do some examples. For a straight line $y = mx + b$, then $f(x) = mx + b$. To find $f(x + h)$ we replace x with $x + h$ to obtain $f(x + h) = m(x + h) + b$. Therefore, the secant slope is equal to

$$\frac{f(x + h) - f(x)}{h} = \frac{m(x + h) + b - (mx + b)}{h} = \frac{mh}{h} = m$$

The tangent slope is also equal to m regardless of the value of x involved, so $f'(x) = m$. This makes sense because the line $y = mx + b$ always has a slope of m.

Let's now find the derivative $y = x^2$ using the definition. Here, we have

$$
\begin{aligned}
\frac{f(x+h) - f(x)}{h} &= \frac{(x+h)^2 - x^2}{h} \\
&= \frac{(x^2 + 2xh + h^2) - x^2}{h} \\
&= \frac{2xh + h^2}{h} \\
&= 2x + h
\end{aligned}
$$

and as h goes to 0, we get $f'(x) = 2x$.

For $f(x) = x^3$, we have

$$
\begin{aligned}
\frac{f(x+h) - f(x)}{h} &= \frac{(x+h)^3 - x^3}{h} \\
&= \frac{(x^3 + 3x^2h + 3xh^2 + h^3) - x^3}{h} \\
&= \frac{3x^2h + 3xh^2 + h^3}{h} \\
&= 3x^2 + 3xh + h^2
\end{aligned}
$$

and as h goes to 0, we get $f'(x) = 3x^2$.

Given the function $y = f(x)$, the process of determining the derivative function $f'(x)$ is called *differentiation*. The good news is that once we have found the derivatives of a few simple functions, we can determine the derivatives of more complicated functions with very little difficulty and without needing to use the formal limit-based definition given above. The following theorem is very useful.

Theorem: If $u(x) = f(x) + g(x)$, then $u'(x) = f'(x) + g'(x)$. In other words, *the derivative of the sum is the sum of the derivatives*. Also, if c is any real number, the derivative of $cf(x)$ is $cf'(x)$.

As a consequence of this theorem, since $y = x^3$ has derivative $3x^2$ and $y = x^2$ has derivative $2x$, then $y = x^3 + x^2$ has derivative $3x^2 + 2x$. To illustrate the second statement, the function $y = 10x^3$ has derivative $30x^2$.

✂ Aside

Proof: Let $u(x) = f(x) + g(x)$. Then

$$\frac{u(x+h) - u(x)}{h} = \frac{f(x+h) + g(x+h) - (f(x) + g(x))}{h}$$

$$= \frac{f(x+h) - f(x)}{h} + \frac{g(x+h) - g(x)}{h}$$

Taking the limit of both sides as $h \to 0$ gives us

$$u'(x) = f'(x) + g'(x) \qquad \square$$

Note that when we take the limit on the right side of the equation, we are using the fact that *the limit of the sum is the sum of the limits*. We won't rigorously prove that here, but the intuition is that if the number a is getting closer and closer to A and b is getting closer and closer to B, then $a + b$ is getting closer and closer to $A + B$. We note that it's also true that *the limit of the product is the product of the limits* and *the limit of the quotient is the quotient of the limits*. But as we'll see, the corresponding rules for derivatives are not quite as straightforward. For instance, the derivative of the product is *not* the product of the derivatives.

For the second half of the theorem, if $v(x) = cf(x)$, then we have

$$v'(x) = \lim_{h \to 0} \frac{v(x+h) - v(x)}{h} = \lim_{h \to 0} \frac{cf(x+h) - cf(x)}{h}$$

$$= c \lim_{h \to 0} \frac{f(x+h) - f(x)}{h} = cf'(x)$$

as desired. $\qquad \square$

To differentiate $f(x) = x^4$, let's first expand $f(x+h) = (x+h)^4 = x^4 + 4x^3h + 6x^2h^2 + 4xh^3 + h^4$. The coefficients of this expression, 1, 4, 6, 4, 1, might look familiar to you, as they are row 4 of Pascal's triangle, studied in Chapter 4. Thus, we have

$$\frac{f(x+h) - f(x)}{h} = \frac{4x^3h + 6x^2h^2 + 4xh^3 + h^4}{h} = 4x^3 + h \times [\text{STUFF}]$$

and as $h \to 0$, we get $f'(x) = 4x^3$. Do you see a pattern? The derivatives of x, x^2, x^3, and x^4 are, respectively, 1, $2x$, $3x^2$, and $4x^3$. Applying the same logic to higher exponents gives us the following *powerful* rule. Another popular notation for the derivative is y', so let's start using it here.

Theorem (the power rule): For $n \geq 0$,

$$y = x^n \text{ has derivative } y' = nx^{n-1}$$

For example,

$$\text{if } y = x^5, \text{ then } y' = 5x^4$$

and

$$\text{if } y = x^{10}, \text{ then } y' = 10x^9$$

Even a constant function, like $y = 1$, can be differentiated by this rule, since $1 = x^0$ and $y = x^0$ has derivative $0x^{-1} = 0$, for every value of x. This makes sense since the line $y = 1$ is horizontal. As a consequence of the power rule and the previous theorem, we can now differentiate any polynomial. For example, if

$$y = x^{10} + 3x^5 - x^3 - 7x + 2520$$

then

$$y' = 10x^9 + 15x^4 - 3x^2 - 7$$

The power rule is even true when n is not a positive integer. For instance, if

$$y = \frac{1}{x} = x^{-1}$$

then

$$y' = -1x^{-2} = \frac{-1}{x^2}$$

Likewise, if

$$y = \sqrt{x} = x^{1/2}$$

then

$$y' = \frac{1}{2}x^{-1/2} = \frac{1}{2\sqrt{x}}$$

But we're not ready to prove these facts yet. Before we learn how to differentiate more complicated functions, let's take advantage of what we have learned so far to solve some other interesting and practical optimization problems.

Max-Min Problems

Differentiation helps us determine where a function achieves its maximum or minimum values. For instance, for what value of x does the parabola $y = x^2 - 8x + 10$ reach its lowest point?

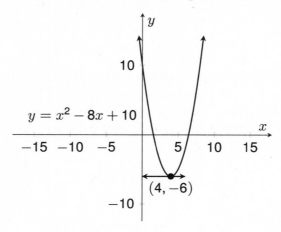

The parabola $y = x^2 - 8x + 10$ is at its lowest point when $y' = 0$

At the lowest point, the slope of the tangent line must be 0. Since $y' = 2x - 8$, solving $2x - 8 = 0$ tells us that the minimum occurs when $x = 4$ (and $y = 16 - 32 + 10 = -6$). For the function $y = f(x)$, a value of x that satisfies $f'(x) = 0$ is called a *critical point* of f. For the function $y = x^2 - 8x + 10$, the only critical point is $x = 4$.

Where does the maximum occur? In the above problem, there is no maximum because the y-value of $x^2 - 8x + 10$ can get arbitrarily big. But if x were restricted to an interval, say $0 \leq x \leq 6$, then y would be greatest at one of its endpoints. Here, we see that when $x = 0$, $y = 10$ and when $x = 6$, $y = -2$, so the function would be maximized at the endpoint $x = 0$. In general, we have the following important theorem.

Theorem (optimization theorem): If a differentiable function $y = f(x)$ is maximized or minimized at a point x^*, then x^* must be either a critical point of f or an endpoint.

Let's return to the box problem at the beginning of the chapter. Here we are interested in maximizing the function

$$y = (12 - 2x)^2 x = 4x^3 - 48x^2 + 144x$$

where x is required to be between 0 and 6. We wish to find a value of x for which y is maximized. Since our function is a polynomial, we see that it has derivative

$$y' = 12x^2 - 96x + 144 = 12(x^2 - 8x + 12) = 12(x-2)(x-6)$$

Hence the function has critical points $x = 2$ and $x = 6$.

The box has volume 0 at the endpoints $x = 0$ and $x = 6$, so the volume is minimized there. It has maximum volume at the other critical point $x = 2$, where $y = 128$ cubic inches.

Differentiation Rules

The more functions that we can differentiate, the more problems we can solve. Perhaps the most important function in calculus is the *exponential function* $y = e^x$. What makes $y = e^x$ special is that it is equal to its own derivative.

Theorem: If $y = e^x$, then $y' = e^x$.

> ✕ **Aside**
>
> Why does $f(x) = e^x$ satisfy $f'(x) = e^x$? Here is the essential idea. First notice that
>
> $$\frac{f(x+h) - f(x)}{h} = \frac{e^{x+h} - e^x}{h} = \frac{e^x(e^h - 1)}{h}$$
>
> Now recall that the number e has definition
>
> $$e = \lim_{n \to \infty} \left(1 + \frac{1}{n}\right)^n$$
>
> which means that as n gets larger and larger, $(1 + 1/n)^n$ gets closer and closer to e. Now let $h = 1/n$. When n is very large, then $h = 1/n$ is very close to 0. Thus for h near 0,
>
> $$e \approx (1 + h)^{1/h}$$
>
> If we raise both sides to the power h and use the exponent law $\left(a^b\right)^c = a^{bc}$, then we see that
>
> $$e^h \approx 1 + h$$
>
> and therefore
>
> $$\frac{e^h - 1}{h} \approx 1$$
>
> Thus as h gets closer and closer to 0, $\frac{e^h - 1}{h}$ gets closer and closer to 1, so $\frac{f(x+h) - f(x)}{h}$ gets closer and closer to e^x.   □

Are there any other functions that are their own derivative? Yes, but they are all of the form $y = ce^x$ where c is a real number. (Note that this includes the case when $c = 0$, giving us the constant function $y = 0$.)

We have seen that when we add functions, the derivative of the sum is the sum of the derivatives. What about the product of functions? Alas, the derivative of a product is not the product of the derivatives, but it's not too hard to compute, as the following theorem demonstrates.

Theorem (product rule for derivatives): If $y = f(x)g(x)$, then

$$y' = f(x)g'(x) + f'(x)g(x)$$

For example, according to the product rule, to differentiate $y = x^3 e^x$, we let $f(x) = x^3$ and $g(x) = e^x$. Therefore

$$\begin{aligned} y' &= f(x)g'(x) + f'(x)g(x) \\ &= x^3 e^x + 3x^2 e^x \end{aligned}$$

Notice that when $f(x) = x^3$ and $g(x) = x^5$, then the product rule says that their product $x^3 x^5 = x^8$ has derivative

$$\begin{aligned} y' &= x^3(5x^4) + 3x^2(x^5) \\ &= 5x^7 + 3x^7 = 8x^7 \end{aligned}$$

which is consistent with the power rule.

✂ Aside

Proof (product rule): Let $u(x) = f(x)g(x)$. Then

$$\frac{u(x+h) - u(x)}{h} = \frac{f(x+h)g(x+h) - f(x)g(x)}{h}$$

Next we cleverly add 0 to the numerator by subtracting and adding $f(x+h)g(x)$ to get

$$\frac{f(x+h)g(x+h) - f(x+h)g(x) + f(x+h)g(x) - f(x)g(x)}{h}$$

$$= f(x+h)\left(\frac{g(x+h) - g(x)}{h}\right) + \left(\frac{f(x+h) - f(x)}{h}\right)g(x)$$

As $h \to 0$, this becomes $f(x)g'(x) + f'(x)g(x)$, as desired.   □

The product rule is not just computationally useful: it also allows us to find derivatives of other functions. For example, we previously proved the power rule for positive exponents. But now we can prove that it is true for fractional and negative exponents too.

For example, the power rule predicts that

$$\text{if } y = \sqrt{x} = x^{1/2}, \text{ then } y' = \frac{1}{2}x^{-1/2} = \frac{1}{2\sqrt{x}}$$

Let's see why this is true using the product rule. Suppose $u(x) = \sqrt{x}$. Then

$$u(x)u(x) = \sqrt{x}\sqrt{x} = x$$

When we differentiate both sides, the product rule tells us that

$$u(x)u'(x) + u'(x)u(x) = 1$$

Thus $2u(x)u'(x) = 1$, and therefore $u'(x) = \frac{1}{2u(x)} = \frac{1}{2\sqrt{x}}$, as predicted.

✂ **Aside**

The power rule also predicts that for negative exponents, $y = x^{-n}$ should have the derivative $y' = -nx^{-n-1} = \frac{-n}{x^{n+1}}$. To prove this, let $u(x) = x^{-n}$, where $n \geq 1$. Then, by definition, for $x \neq 0$, we have

$$u(x)x^n = x^{-n}x^n = x^0 = 1$$

When we differentiate both sides, the product rule tells us that

$$u(x)(nx^{n-1}) + u'(x)x^n = 0$$

Dividing through by x^n and moving the first term to the other side, we get

$$u'(x) = -n\frac{u(x)}{x} = \frac{-n}{x^{n+1}}$$

as desired.   □

Thus, if $y = 1/x = x^{-1}$, then $y' = -1/x^2$.
If $y = 1/x^2 = x^{-2}$, then $y' = -2x^{-3} = -2/x^3$, and so on.

In Chapter 7 we wanted to find the positive number x that would minimize the function

$$y = x + 1/x$$

Using clever geometry, we showed that this occurs when $x = 1$. But with calculus, we don't have to be as clever. We simply solve $y' = 0$, which gives us $1 - 1/x^2 = 0$, and the only positive number that satisfies this is $x = 1$.

The trigonometric functions are also easy to differentiate. Note that for the following theorem to be true, the angles must be expressed in radians.

Theorem: If $y = \sin x$, then $y' = \cos x$. If $y = \cos x$, then $y' = -\sin x$. In other words, *the derivative of sine is cosine* and *the derivative of cosine is negative sine.*

⚔ Aside

Proof: The proof relies on the following lemma. (A lemma is a statement that helps us prove a more important theorem.)

Lemma:

$$\lim_{h \to 0} \frac{\sin h}{h} = 1 \quad \text{and} \quad \lim_{h \to 0} \frac{\cos h - 1}{h} = 0$$

This says that for any tiny angle h (in radians) near 0, its sine value is very close to h and its cosine value is very close to 1. For example, a calculator reveals that $\sin 0.0123 = 0.0122996\ldots$ and $\cos 0.0123 = 0.9999243\ldots$. Assuming the lemma for now, we can differentiate the sine and cosine functions. Using the $\sin(A + B)$ identity from Chapter 9, we have

$$\frac{\sin(x + h) - \sin x}{h} = \frac{\sin x \cos h + \sin h \cos x - \sin x}{h}$$

$$= \sin x \left(\frac{\cos h - 1}{h} \right) + \cos x \left(\frac{\sin h}{h} \right)$$

As $h \to 0$, then according to our lemma, the expression above becomes $(\sin x)(0) + (\cos x)(1) = \cos x$. Likewise,

$$\frac{\cos(x + h) - \cos x}{h} = \frac{\cos x \cos h - \sin x \sin h - \cos x}{h}$$

$$= \cos x \left(\frac{\cos h - 1}{h} \right) - \sin x \left(\frac{\sin h}{h} \right)$$

As $h \to 0$, this becomes $(\cos x)(0) - (\sin x)(1) = -\sin x$, as desired. □

✕ Aside

We can prove $\lim_{h \to 0} \frac{\sin h}{h} = 1$ using the figure below.

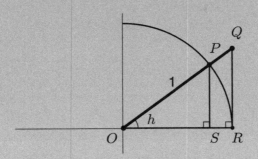

On the unit circle above, $R = (1,0)$ and $P = (\cos h, \sin h)$, where h is a small positive angle. Also, in right triangle OQR,

$$\tan h = \frac{QR}{OR} = \frac{QR}{1} = QR$$

It follows that right triangle OPS has area $\frac{1}{2} \cos h \sin h$, and right triangle OQR has area $\frac{1}{2}OR QR = \frac{1}{2} \tan h = \frac{\sin h}{2 \cos h}$.

Now focus on the sector OPR, which is a wedge-shaped object. The area of the unit circle is $\pi 1^2 = \pi$, and sector OPS is just a fraction $h/(2\pi)$ of the unit circle. Therefore sector OPR has area $\pi(h/2\pi) = h/2$.

Since sector OPR contains triangle OPS and is contained inside triangle OQR, then by comparing their areas we have

$$\frac{1}{2} \cos h \sin h < \frac{h}{2} < \frac{\sin h}{2 \cos h}$$

Multiplying through by $\frac{2}{\sin h} > 0$, we have

$$\cos h < \frac{h}{\sin h} < \frac{1}{\cos h}$$

For positive numbers, if $a < b < c$, then $1/c < 1/b < 1/a$. Therefore,

$$\cos h < \frac{\sin h}{h} < \frac{1}{\cos h}$$

Now as $h \to 0$, both $\cos h$ and $1/\cos h$ go to 1, as desired. Therefore, $\lim_{h \to 0} \frac{\sin h}{h} = 1$. $\qquad \square$

> ✂ **Aside**
>
> We can prove $\lim_{h \to 0} \frac{\cos h - 1}{h} = 0$ using the previous result and a few lines of algebra (including $\cos^2 h + \sin^2 h = 1$).
>
> $$\frac{\cos h - 1}{h} = \frac{\cos h - 1}{h} \cdot \frac{\cos h + 1}{\cos h + 1} = \frac{\cos^2 h - 1}{h(\cos h + 1)}$$
>
> $$= \frac{-\sin^2 h}{h(\cos h + 1)} = -\frac{\sin h}{h} \cdot \frac{\sin h}{\cos h + 1}$$
>
> Now as $h \to 0$, $\frac{\sin h}{h} \to 1$, and $\frac{\sin h}{\cos h + 1} \to \frac{0}{2} = 0$.
> Hence $\lim_{h \to 0} \frac{\cos h - 1}{h} = 0$.   □

Once we know the derivatives of sine and cosine, we can differentiate the tangent function.

Theorem: For $y = \tan x$, $y' = 1/(\cos^2 x) = \sec^2 x$.

Proof: Let $u(x) = \tan x = (\sin x)/(\cos x)$. Then

$$\tan(x) \cos x = \sin x$$

Differentiating both sides and using the product rule, we have

$$\tan x (-\sin x) + \tan'(x) \cos x = \cos x$$

Dividing through by $\cos x$ and solving for $\tan'(x)$ gives us

$$\tan'(x) = 1 + \tan x \tan x = 1 + \tan^2 x = \frac{1}{\cos^2 x} = \sec^2 x$$

where the second-to-last equality is obtained by dividing the identity $\cos^2 x + \sin^2 x = 1$ by $\cos^2 x$.   □

Taking a similar approach allows us to prove the *quotient rule* for differentiation.

Theorem (quotient rule): If $u(x) = f(x)/g(x)$, then

$$u'(x) = \frac{g(x)f'(x) - f(x)g'(x)}{g(x)g(x)}$$

✂ **Aside**

Proof of quotient rule: Since $u(x)g(x) = f(x)$, then when we differentiate both sides, the product rule gives us

$$u(x)g'(x) + u'(x)g(x) = f'(x)$$

If we multiply both sides by $g(x)$, we get

$$g(x)u(x)g'(x) + u'(x)g(x)g(x) = g(x)f'(x)$$

Replacing $g(x)u(x)$ with $f(x)$ and solving for $u'(x)$ gives us the desired quantity.   □

We know how to differentiate polynomials, exponential functions, trigonometric functions, and more. We have seen how to differentiate functions when they are added, multiplied, and divided. The **chain rule** (stated below, but not proved) tells us what to do when the functions are *composed*. For example, if $f(x) = \sin x$ and $g(x) = x^3$, then

$$f(g(x)) = \sin(g(x)) = \sin(x^3)$$

Note that this is not the same as the function

$$g(f(x)) = g(\sin x) = (\sin x)^3$$

Theorem (chain rule): If $y = f(g(x))$, then $y' = f'(g(x))g'(x)$.

For example, if $f(x) = \sin x$, and $g(x) = x^3$, then $f'(x) = \cos x$ and $g'(x) = 3x^2$. The chain rule tells us that if $y = f(g(x)) = \sin(x^3)$, then

$$y' = f'(g(x))g'(x) = \cos(g(x))g'(x) = 3x^2 \cos(x^3)$$

More generally, the chain rule tells us that if $y = \sin(g(x))$, then $y' = g'(x)\cos(g(x))$. By the same logic, $y = \cos(g(x))$ has $y' = -g'(x)\sin(g(x))$.

On the other hand, for the function $y = g(f(x)) = (\sin x)^3$, the chain rule says that

$$y' = g'(f(x))f'(x) = 3(f(x)^2)f'(x) = 3\sin^2 x \cos x$$

More generally, the chain rule tells us that if $y = (g(x))^n$, then $y' = n(g(x))^{n-1}g'(x)$. What does it say about differentiating $y = (x^3)^5$?

$$y' = 5(x^3)^4(3x^2) = 5x^{12}(3x^2) = 15x^{14}$$

which is consistent with the power rule.

Let's differentiate $y = \sqrt{x^2 + 1} = (x^2 + 1)^{1/2}$. Then

$$y' = \frac{1}{2}(x^2 + 1)^{-1/2}(2x) = \frac{x}{\sqrt{x^2 + 1}}$$

Exponential functions are easy to differentiate too. Since e^x is its own derivative, if $y = e^{g(x)}$, then

$$y' = g'(x)e^{g(x)}$$

For example, $y = e^{x^3}$ has $y' = (3x^2)e^{x^3}$.

Notice that the function $y = e^{kx}$ has derivative $y' = ke^{kx} = ky$. This is one of the properties that makes exponential functions so important. They arise anytime that the rate of growth of a function is proportional to the size of its output. This is why exponential functions arise so often in financial and biological problems.

The *natural logarithm* function $\ln x$ has the property that

$$e^{\ln x} = x$$

for any $x > 0$. Let's use the chain rule to determine its derivative. Letting $u(x) = \ln x$, we have $e^{u(x)} = x$. Differentiating both sides of this equation tells us $u'(x)e^{u(x)} = 1$. But since $e^{u(x)} = x$, it follows that $u'(x) = 1/x$. In other words, if $y = \ln x$, then $y' = 1/x$. Applying the chain rule again, we obtain that if $y = \ln(g(x))$, then $y' = \frac{g'(x)}{g(x)}$.

We summarize these consequences of the chain rule here.

$y = f(g(x))$	$y' = f'(g(x))g'(x)$
$y = \sin(g(x))$	$y' = g'(x)\cos(g(x))$
$y = \cos(g(x))$	$y' = -g'(x)\sin(g(x))$
$y = (g(x))^n$	$y' = n(g(x))^{n-1}g'(x)$
$y = e^{g(x)}$	$y' = g'(x)e^{g(x)}$
$y = \ln(g(x))$	$y' = g'(x)/g(x)$

Let's apply the chain rule to solve a problem from cow-culus! Clara the Cow is one mile north of the x-axis river, which runs from east to west. Her barn is three miles east and one mile north of her current position. She wishes to drink from the river and then walk to her barn so as to minimize her total amount of walking. Where on the river should she stop to drink?

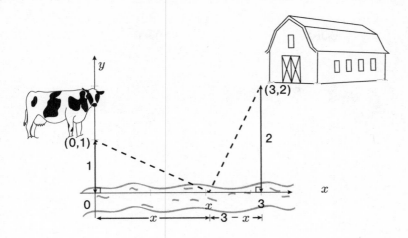

Cow-culus question: Where should the cow drink to minimize the total amount of walking?

Assuming that Clara walks in a straight line from her starting point $(0,1)$ to the drinking point $(x,0)$, then the Pythagorean theorem (or the distance formula) tells us that the length of the line to the drinking point is $\sqrt{x^2+1}$ and the length of the trip to the barn at $B = (3,2)$ is $\sqrt{(3-x)^2+4} = \sqrt{x^2-6x+13}$. Hence the problem is to determine the number x (between 0 and 3) that minimizes

$$y = \sqrt{x^2+1} + \sqrt{x^2-6x+13} = (x^2+1)^{1/2} + (x^2-6x+13)^{1/2}$$

When we differentiate the above expression (using the chain rule) and set it equal to 0, we get

$$\frac{x}{\sqrt{x^2+1}} + \frac{x-3}{\sqrt{x^2-6x+13}} = 0$$

You can verify that when $x = 1$, the left side of the above equation becomes $1/\sqrt{2} - 2/\sqrt{8}$, which indeed equals 0. (You can solve the equation directly by putting $x/\sqrt{x^2+1}$ on the other side of the equation, then squaring both sides and cross-multiplying. A lot of cancellation occurs, and the only solution between 0 and 3 is $x = 1$.)

We can confirm our answer with just a little *reflection* like we did in Chapter 7. Imagine that instead of Clara going to the barn at $(3,2)$ after her drink, she went to the barn's reflection at $B' = (3,-2)$, as in the figure on the next page.

The distance to B' is exactly the same as the distance to B. And every point from above the river to below the river must cross the x-axis somewhere. The path with the shortest distance is the straight line

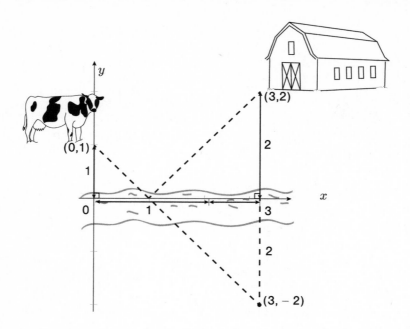

Upon reflection, there is another way to solve this problem

from $(0,1)$ to $(3,-2)$ (with slope $-3/3 = -1$), which intersects the x-axis when $x = 1$. No calculus or square roots required!

A Magical Application: Taylor Series

When we proved Euler's equation at the end of last chapter, we relied on the following mysterious formulas:

$$e^x = 1 + x + \frac{x^2}{2!} + \frac{x^3}{3!} + \frac{x^4}{4!} + \cdots$$

$$\cos x = 1 - \frac{x^2}{2!} + \frac{x^4}{4!} - \frac{x^6}{6!} + \cdots$$

$$\sin x = x - \frac{x^3}{3!} + \frac{x^5}{5!} - \frac{x^7}{7!} + \cdots$$

Before we see how we arrived at these, let's play with them for a little bit. Look what happens as you differentiate every term in the series for e^x. For example, the power rule tells us that the derivative of $x^4/4!$ is $(4x^3)/4! = x^3/3!$, which is the preceding term of the series. In other words, when we differentiate the series for e^x, we get back the series for e^x, which agrees with what we know about e^x!

If we differentiate $x - x^3/3! + x^5/5! - x^7/7! + \cdots$ term by term, we get $1 - x^2/2! + x^4/4! - x^6/6! + \cdots$, which is consistent with the fact that the derivative of the sine function is the cosine function. Likewise, when we differentiate the cosine series we get the negative of the sine series. Notice also that the series confirms that $\cos 0 = 1$, and because every exponent is even, the value of $\cos(-x)$ will be the same as $\cos x$, which we know to be true. (For example, $(-x)^4/4! = x^4/4!$.) Likewise, for the sine series, we see that $\sin 0 = 0$ and because all of the exponents are odd, we get $\sin(-x) = -\sin x$, as appropriate.

Now let's see where those formulas come from. In this chapter, we have learned how to find the derivative of most commonly used functions. But there are times when it is useful to differentiate a function more than once by computing its second derivative or third derivative or more, denoted by $f''(x)$, $f'''(x)$, and so on. The second derivative $f''(x)$ measures the rate of change of the slope of the function (also known as its *concavity*) at the point $(x, f(x))$. The third derivative measures the rate of change of the second derivative, and so on.

The formulas given above are called *Taylor series*, named after English mathematician Brook Taylor (1685–1731). For a function $f(x)$ with derivatives $f'(x)$, $f''(x)$, $f'''(x)$, and so on, we have

$$f(x) = f(0) + f'(0)x + f''(0)\frac{x^2}{2!} + f'''(0)\frac{x^3}{3!} + f''''(0)\frac{x^4}{4!} + \cdots$$

for all values of x that are "close enough" to 0. What does close enough mean? For some functions, like e^x, $\sin x$, and $\cos x$, all values of x are close enough. But for some functions, as we'll see later, x has to be small for the series to match the function.

Let's see what the formula says for $f(x) = e^x$. Since e^x is its own first (and second and third ...) derivative, it follows that

$$f(0) = f'(0) = f''(0) = f'''(0) = \cdots = e^0 = 1$$

so the Taylor series for e^x is $1 + x + x^2/2! + x^3/3! + x^4/4! + \ldots$, as promised. When x is small, then we only need to compute a few terms of the series to get an excellent approximation of the exact answer.

Let's apply this to compound interest. In the last chapter we saw that if we have $1000 earning 5 percent interest, compounded continuously, then at the end of the year we have $1000e^{0.05} = \$1051.27$. But we can get a get a good estimate of that with the *second-order Taylor polynomial approximation*

$$\$1000(1 + 0.05 + (0.05)^2/2!) = \$1051.25$$

and the third-order approximation gives us $1051.27.

We illustrate Taylor approximation in the figure below, where $y = e^x$ is graphed along with its first three Taylor polynomials.

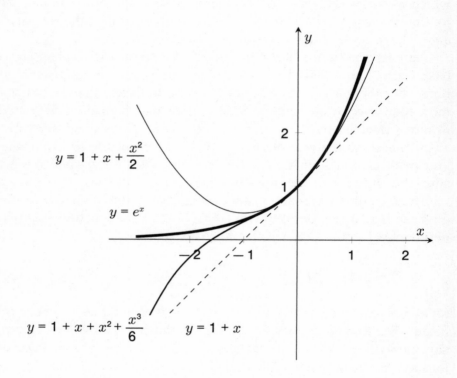

Taylor approximations of e^x

As we increase the degree of the Taylor polynomial, the approximation becomes more and more accurate, especially for the values of x near 0. So what is it about the Taylor polynomials that make them work so well? The first-degree approximation (also called the *linear approximation*) says that for x near 0,

$$f(x) \approx f(0) + f'(0)x$$

This is a straight line that goes through the point $(0, f(0))$ and has slope $f'(0)$. Likewise, it can be shown that the nth-degree Taylor polynomial goes through the point $(0, f(0))$ with the same first derivative, same second derivative, same third derivative, and so on, up to the same nth derivative as the original function $f(x)$.

> ✗ **Aside**
>
> Taylor polynomials and Taylor series can also be defined for values of x near other numbers besides 0. Specifically, the Taylor series for $f(x)$ with *basepoint* a is equal to
>
> $$f(a) + f'(a)(x - a) + f''(a)\frac{(x - a)^2}{2!} + f'''(a)\frac{(x - a)^3}{3!} + \cdots$$
>
> As in the case when $a = 0$, the Taylor series is equal to $f(x)$ for all real or complex numbers x sufficiently close to a.

Let's look at the Taylor series for $f(x) = \sin x$. Notice that $f'(x) = \cos x$, $f''(x) = -\sin x$, $f'''(x) = -\cos x$, and $f''''(x) = \sin x = f(x)$ again. When this is evaluated at 0, starting at $f(0)$, we obtain the cyclic pattern $0, 1, 0, -1, 0, 1, 0, -1, \ldots$, causing every even power of x to disappear in the Taylor series. Thus we have, for all values of x (measured in radians),

$$\sin x = x - \frac{x^3}{3!} + \frac{x^5}{5!} - \frac{x^7}{7!} + \cdots$$

And similarly, when $f(x) = \cos x$ we get

$$\cos x = 1 - \frac{x^2}{2!} + \frac{x^4}{4!} - \frac{x^6}{6!} + \cdots$$

Finally, let's look at an example where the Taylor series is equal to the function for some values of x, but not all values of x. Consider $f(x) = \frac{1}{1-x} = (1-x)^{-1}$. Here $f(0) = 1$ and using the chain rule, the first few derivatives are

$$f'(x) = -1(1-x)^{-2}(-1) = (1-x)^{-2}$$

$$f''(x) = (-2)(1-x)^{-3}(-1) = 2(1-x)^{-3}$$

$$f'''(x) = -6(1-x)^{-4}(-1) = 3!(1-x)^{-4}$$

$$f''''(x) = -4!(1-x)^{-5}(-1) = 4!(1-x)^{-5}$$

Continuing this pattern (or using a proof by induction), we see that the nth derivative of $(1 - x)^{-1}$ is $n!(1 - x)^{-(n+1)}$, and when $x = 0$, the nth derivative is just $n!$. Consequently, the Taylor series gives us

$$\frac{1}{1 - x} = 1 + x + x^2 + x^3 + x^4 + \cdots$$

but this equation is only valid when x is between -1 and 1. For example, when x is greater than 1, then the numbers being added get larger and larger and so the sum is undefined.

We will say more about this series in the next chapter. In the meantime, you might wonder what it really means to add up an infinite number of numbers. How could such a sum be equal to anything? That's a fair question, and we will attempt to answer it as we explore the nature of infinity, where we will encounter many surprising, puzzling, unintuitive, and beautiful results.

CHAPTER TWELVE

$1 + 2 + 3 + \cdots = \infty$

The Magic of Infinity

Infinitely Interesting

Last, but certainly not least, let's talk about infinity. Our journey began in Chapter 1 with the sum of the numbers from 1 to 100:

$$1 + 2 + 3 + 4 + \cdots + 100 = 5050$$

and we eventually discovered formulas for the sum of the numbers from 1 to n:

$$1 + 2 + 3 + \cdots + n = \frac{n(n+1)}{2}$$

and we discovered formulas for other sums with a finite number of terms. In this chapter, we will explore sums that have an infinite number of terms like

$$1 + \frac{1}{2} + \frac{1}{4} + \frac{1}{8} + \frac{1}{16} + \cdots$$

which, I hope, I will convince you has a sum that is equal to 2. Not *approximately* 2, but *exactly* equal to 2. Some sums have intriguing answers, like

$$1 - \frac{1}{3} + \frac{1}{5} - \frac{1}{7} + \frac{1}{9} - \frac{1}{11} + \cdots = \frac{\pi}{4}$$

And some infinite sums, like

$$1 + \frac{1}{2} + \frac{1}{3} + \frac{1}{4} + \frac{1}{5} + \frac{1}{6} + \cdots$$

279

don't add up to anything. We say that the sum of all the positive numbers is *infinity*, denoted

$$1 + 2 + 3 + 4 + 5 + \cdots = \infty$$

which means that the sum grows without bound. In other words, the sum will eventually exceed any number you wish: it will eventually exceed one hundred, then one million, then one quadrillion, and so on. And yet, by the end of this chapter, we shall see that a case could be made that

$$1 + 2 + 3 + 4 + 5 + \cdots = \frac{-1}{12}$$

Are you intrigued? I hope so! As we shall see, when you enter the twilight zone of infinity, very strange things can happen, which is part of what makes mathematics so fascinating and fun.

Is infinity a number? Not really, although it sometimes gets treated that way. Loosely speaking, mathematicians might say:

$$\infty + 1 = \infty \qquad \infty + \infty = \infty \qquad 5 \times \infty = \infty \qquad \frac{1}{\infty} = 0$$

Technically, there is no largest number, since we can always add 1 to obtain a larger number. The symbol ∞ essentially means "arbitrarily large" or bigger than any positive number. Likewise, the term $-\infty$ means less than any negative number. By the way, the quantities $\infty - \infty$ (infinity minus infinity) and $1/0$ are undefined. It is tempting to define $1/0 = \infty$, since when we divide 1 by smaller and smaller positive numbers, the quotient gets bigger and bigger. But the problem is that when we divide 1 by tiny negative numbers, the quotient gets more and more negative.

An Important Infinite Sum: The Geometric Series

Let's start with a statement that is accepted by all mathematicians, but seems wrong to most people when they first see it:

$$0.99999\ldots = 1$$

Everyone agrees that the two numbers are close, indeed extremely close, but many still feel that they should not be considered the same number. Let me try to convince you that the numbers are in fact equal by offering various different proofs. I hope that at least one of these explanations will satisfy you.

Perhaps the quickest proof is that if you accept the statement that

$$\frac{1}{3} = 0.33333\ldots$$

then when you multiply both sides by 3 you get

$$1 = \frac{3}{3} = 0.99999\ldots$$

Another proof is to use the technique that we used in Chapter 6 to evaluate repeating decimals. Let's denote the infinite decimal expansion with the variable w as follows:

$$w = 0.99999\ldots$$

Now if we multiply both sides by 10, then we get

$$10w = 9.99999\ldots$$

Subtracting the first equation from the second gives us

$$9w = 9.00000\ldots$$

which means that $w = 1$.

Here's an argument that uses no algebra at all. Do you agree that if two numbers are different, then there must be a different number in between them (for instance, their average)? Then suppose, to the contrary, that $0.99999\ldots$ and 1 were different numbers. If that were the case, what number would be in between them? If you can't find another number between them, then they can't be different numbers.

We say that two numbers or infinite sums are *equal* if they are *arbitrarily close* to one another. In other words, the difference between the two quantities is less than any positive number you can name, whether it be 0.01 or 0.0000001 or 1 divided by a trillion. Since the difference between 1 and $0.99999\ldots$ is smaller than any positive number, then mathematicians agree to call these quantities equal.

It's with the same logic that we can evaluate the infinite sum below:

$$1 + \frac{1}{2} + \frac{1}{4} + \frac{1}{8} + \frac{1}{16} + \cdots = 2$$

We can give this sum a physical interpretation. Imagine you are standing two meters away from a wall, and you take one big step exactly one meter toward the wall, then another step half a meter toward the wall, then a quarter of a meter, then an eighth of a meter, and so on. After each step, the distance between you and the wall is cut exactly in half. Ignoring the practical limitations of taking tinier and tinier steps, you eventually get as close to the wall as desired. Hence the total length of your steps would be exactly two meters.

We can illustrate this sum geometrically, as in the figure below. We start with a 1-by-2 rectangle with area 2, then cut it in half, then half again, then half again, and so on. The area of the first region is 1. The next region has area 1/2, then the next region has area 1/4, and so on. As n goes to infinity, the regions fill up the entire rectangle, and so their total area is 2.

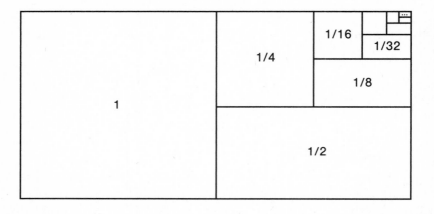

A geometric proof that $1 + 1/2 + 1/4 + 1/8 + 1/16 + \cdots = 2$

For a more algebraic explanation, we look at the *partial sums*, as given in the table below.

Partial sums of $1 + \frac{1}{2} + \frac{1}{4} + \frac{1}{8} + \cdots$		
1	$= 1$	$= 2 - 1$
$1 + \frac{1}{2}$	$= 1\frac{1}{2}$	$= 2 - \frac{1}{2}$
$1 + \frac{1}{2} + \frac{1}{4}$	$= 1\frac{3}{4}$	$= 2 - \frac{1}{4}$
$1 + \frac{1}{2} + \frac{1}{4} + \frac{1}{8}$	$= 1\frac{7}{8}$	$= 2 - \frac{1}{8}$
$1 + \frac{1}{2} + \frac{1}{4} + \frac{1}{8} + \frac{1}{16}$	$= 1\frac{15}{16}$	$= 2 - \frac{1}{16}$
$1 + \frac{1}{2} + \frac{1}{4} + \frac{1}{8} + \frac{1}{16} + \frac{1}{32}$	$= 1\frac{31}{32}$	$= 2 - \frac{1}{32}$
$\vdots$	$\vdots$	$\vdots$

The pattern seems to indicate that for $n \geq 0$,

$$1 + \frac{1}{2} + \frac{1}{4} + \frac{1}{8} + \cdots + \frac{1}{2^n} = 2 - \frac{1}{2^n}$$

We can prove this by induction (as we learned in Chapter 6) or as a special case of the finite geometric series formula below.

Theorem (finite geometric series): For $x \neq 1$ and $n \geq 0$,

$$1 + x + x^2 + x^3 + \cdots + x^n = \frac{1 - x^{n+1}}{1 - x}$$

Proof 1: This can be proved by induction as follows. When $n = 0$, the formula says that $1 = \frac{1-x^1}{1-x}$, which is certainly true. Now assume the formula holds when $n = k$, so that

$$1 + x + x^2 + x^3 + \cdots + x^k = \frac{1 - x^{k+1}}{1 - x}$$

Then the formula will continue to be true when $n = k + 1$, since when we add x^{k+1} to both sides, we get

$$
\begin{aligned}
1 + x + x^2 + x^3 + \cdots + x^k + x^{k+1} &= \frac{1 - x^{k+1}}{1 - x} + x^{k+1} \\
&= \frac{1 - x^{k+1}}{1 - x} + \frac{x^{k+1}(1 - x)}{1 - x} \\
&= \frac{1 - x^{k+1} + x^{k+1} - x^{k+2}}{1 - x} \\
&= \frac{1 - x^{k+2}}{1 - x}
\end{aligned}
$$

as desired.   □

Alternatively, we can prove this by *shifty* algebra, as follows.

Proof 2: Let

$$S = 1 + x + x^2 + x^3 + \cdots + x^n$$

Then when we multiply both sides by x we get

$$xS = \quad x + x^2 + x^3 + \cdots + x^n + x^{n+1}$$

Subtracting away the xS (pronounced "excess"), we have massive amounts of cancellation, leaving us with

$$S - xS = 1 - x^{n+1}$$

In other words, $S(1 - x) = 1 - x^{n+1}$, and therefore

$$S = \frac{1 - x^{n+1}}{1 - x}$$

as desired.   □

Notice that when $x = 1/2$, the finite geometric series confirms our earlier pattern:

$$1 + \frac{1}{2} + \frac{1}{4} + \frac{1}{8} + \cdots + \frac{1}{2^n} = \frac{1 - (1/2)^{n+1}}{1 - \frac{1}{2}} = 2 - \frac{1}{2^n}$$

As n gets larger and larger, $(1/2)^n$ gets closer and closer to 0. Thus, as $n \to \infty$, we have

$$
\begin{aligned}
1 + \frac{1}{2} + \frac{1}{4} + \frac{1}{8} + \frac{1}{16} + \cdots &= \lim_{n \to \infty} \left(1 + \frac{1}{2} + \frac{1}{4} + \frac{1}{8} + \cdots + \frac{1}{2^n} \right) \\
&= \lim_{n \to \infty} \left(2 - \frac{1}{2^n} \right) \\
&= 2
\end{aligned}
$$

> **✂ Aside**
>
> Here's a joke that only mathematicians find funny. An infinite number of mathematicians walk into a bar. The first mathematician says, "I'd like one glass of beer." The second mathematician says, "I'd like half a glass of beer." The third mathematician says, "I'd like a quarter of a glass of beer." The fourth mathematician says, "I'd like an eighth of a glass...." The bartender shouts, "Know your limits!" and hands them two beers.

More generally, any number between -1 and 1 that gets raised to higher and higher powers will get closer and closer to 0. Thus we have the all-important *(Infinite) geometric series*.

Theorem (geometric series): For $-1 < x < 1$,

$$1 + x + x^2 + x^3 + x^4 + \cdots = \frac{1}{1 - x}$$

The geometric series solves the last problem by letting $x = 1/2$:

$$1 + \frac{1}{2} + \frac{1}{4} + \frac{1}{8} + \frac{1}{16} + \cdots = \frac{1}{1 - 1/2} = 2$$

If the geometric series looks familiar, it's because we encountered it at the end of the last chapter when we used calculus to show that the function $y = 1/(1-x)$ has Taylor series $1 + x + x^2 + x^3 + x^4 + \cdots$.

Let's see what else the geometric series tells us. What can we say about the following sum?

$$\frac{1}{4} + \frac{1}{16} + \frac{1}{64} + \frac{1}{256} + \cdots$$

When we factor the number $1/4$ out of each term, this becomes

$$\frac{1}{4}\left(1 + \frac{1}{4} + \frac{1}{16} + \frac{1}{64} + \cdots\right)$$

so the geometric series (with $x = 1/4$) says that this simplifies to

$$\frac{1}{4}\left(\frac{1}{1-1/4}\right) = \frac{1}{4} \times \frac{4}{3} = \frac{1}{3}$$

That series has a particularly beautiful proof without words as shown on the next page. Notice that the dark squares occupy exactly one-third of the area of the big square.

We can even use the geometric series to settle the 0.99999... question, since an infinite decimal expansion is just an infinite series in disguise. Specifically, we can use the geometric series with $x = 1/10$ to get

$$\begin{aligned}
0.99999\ldots &= \frac{9}{10} + \frac{9}{100} + \frac{9}{1000} + \frac{9}{10000} + \cdots \\
&= \frac{9}{10}\left(1 + \frac{1}{10} + \frac{1}{100} + \frac{1}{1000} + \cdots\right) \\
&= \frac{9}{10}\left(\frac{1}{(1-1/10)}\right) \\
&= \frac{9}{10-1} \\
&= 1
\end{aligned}$$

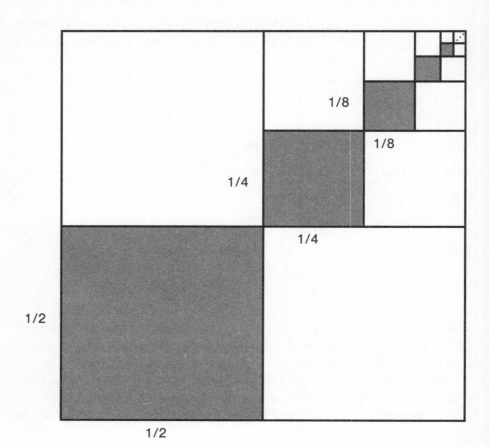

Proof without words: $1/4 + 1/16 + 1/64 + 1/256 + \cdots = 1/3$

The geometric series formula even works when x is a complex number, provided that the length of x is less than 1. For example, the imaginary number $i/2$ has length $1/2$, so the geometric series tells us that

$$1 + i/2 + (i/2)^2 + (i/2)^3 + (i/2)^4 + \cdots = \frac{1}{1 - i/2}$$

$$= \frac{2}{2 - i} = \frac{2}{2 - i} \times \frac{2 + i}{2 + i} = \frac{4 + 2i}{4 - i^2} = \frac{4 + 2i}{5} = \frac{4}{5} + \frac{2}{5}i$$

which we illustrate on the next page on the complex plane.

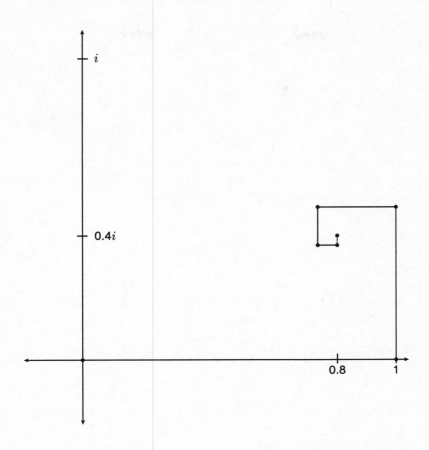

$$1 + i/2 + (i/2)^2 + (i/2)^3 + (i/2)^4 + (i/2)^5 + \cdots = \tfrac{4}{5} + \tfrac{2}{5}i$$

Although the finite geometric series formula is valid for all values of $x \neq 1$, the (infinite) geometric series formula requires that $|x| < 1$. For example, when $x = 2$, the finite geometric series correctly tells us (as we derived in Chapter 6) that

$$1 + 2 + 4 + 8 + 16 + \cdots + 2^n = \frac{1 - 2^{n+1}}{1 - 2} = 2^{n+1} - 1$$

but substituting $x = 2$ in the geometric series formula says that

$$1 + 2 + 4 + 8 + 16 + \cdots = \frac{1}{1 - 2} = -1$$

which looks ridiculous. (Although looks can be deceiving. We will actually see a plausible interpretation of this result in our last section.)

✕ Aside

There are infinitely many positive integers:

$$1, 2, 3, 4, 5 \ldots$$

There are also infinitely many positive even integers:

$$2, 4, 6, 8, 10 \ldots$$

Mathematicians say that the set of positive integers and the set of even positive integers have the same *size* (or cardinality or level of infinity) because they can be paired up with one another:

$$
\begin{array}{ccccc}
1 & 2 & 3 & 4 & 5 \quad \cdots \\
\updownarrow & \updownarrow & \updownarrow & \updownarrow & \updownarrow \quad \cdots \\
2 & 4 & 6 & 8 & 10 \quad \cdots
\end{array}
$$

A set that can be paired up with the positive integers is called *countable*. Countable sets have the smallest level of infinity. Any set that can be *listed* is countable, since the first element in the list is paired up with 1, the second element is paired up with 2, and so on. The set of all integers

$$\ldots -3, -2, -1, 0, 1, 2, 3 \ldots$$

can't be listed from smallest to largest (what would be the first number on the list?), but they can be listed this way:

$$0, 1, -1, 2, -2, 3, -3 \ldots$$

Thus the set of all integers is countable, and of the same size as the number of positive integers.

How about the set of positive rational numbers? These are the numbers of the form m/n where m and n are positive integers. Believe it or not, this set is countable too. They can be listed as follows:

$$\frac{1}{1}, \quad \frac{1}{2}, \frac{2}{1}, \quad \frac{1}{3}, \frac{2}{2}, \frac{3}{1}, \quad \frac{1}{4}, \frac{2}{3}, \frac{3}{2}, \frac{4}{1} \cdots$$

where we first list the fractions according to the sum of their numerators and denominators. Since every rational number appears on the list, the positive rational numbers are countable too.

✕ Aside

Are there any infinite sets of numbers that are *not* countable? The German mathematician Georg Cantor (1845–1918) proved that the real numbers, even when restricted to those that lie between 0 and 1, form an *uncountable* set. You might try to list them this way:

$$0.1, 0.2, \ldots, 0.9, \ 0.01, 0.02, \ldots, 0.99, \ 0.001, 0.002, \ldots 0.999, \ \ldots$$

and so on. But that will only generate real numbers with a finite number of digits. For instance, the number $1/3 = 0.333\ldots$ will never appear on this list. But could there be a more creative way to list all the real numbers? Cantor proved that this would be impossible, by reasoning as follows. Suppose, to the contrary, that the real numbers were listable. To give a concrete example, suppose the list began as

$$0.314159265\ldots$$

$$0.271828459\ldots$$

$$0.618033988\ldots$$

$$0.123581321\ldots$$

$$\vdots$$

We can prove that such a list is guaranteed to be incomplete by creating a real number that won't be on the list. Specifically, we create the real number $0.r_1 r_2 r_3 r_4 \ldots$ where r_1 is an integer between 0 and 9 and differs from the first number in the first digit (in our example, $r_1 \neq 3$) and r_2 differs from the second number in the second digit (here $r_2 \neq 7$) and so on. For instance, we might create the number $0.2674\ldots$. Such a number can't be on the list anywhere. Why is it not the millionth number on the list? Because it differs in the millionth decimal place. Hence any example of a list you create is guaranteed to be missing some numbers, so the real numbers are not countable. This is called Cantor's diagonalization argument, but I like to call it proof by Cantor-example. (Sorry.)

In essence, we have shown that although there are infinitely many rational numbers, there are considerably more irrational numbers. If you randomly choose a real number from the real line, it will almost certainly be irrational.

Infinite series arise frequently in probability problems. Suppose you roll two 6-sided dice repeatedly until a total of 6 or 7 appears. If a 6 occurs before a 7 occurs, then you win the bet. Otherwise, you lose. What is your chance of winning? There are $6 \times 6 = 36$ equally likely dice rolls. Of these, 5 of them have a total of 6 (namely $(1,5)$, $(2,4)$, $(3,3)$, $(4,2)$, $(5,1)$) and 6 of them have a total of 7 $((1,6)$, $(2,5)$, $(3,4)$, $(4,3)$, $(5,2)$, $(6,1))$. Hence it would seem that your chance of winning should be less than 50 percent. Intuitively, as you roll the dice, there are only $5 + 6 = 11$ rolls that matter—all the rest require us to roll again. Of these 11 numbers, 5 of them are winners and 6 are losers. Thus it would seem that your chance of winning should be 5/11.

We can confirm that the probability of winning is indeed 5/11 using the geometric series. The probability of winning on the first roll of the dice is 5/36. What is the chance of winning on the second roll? For this to happen, you must not roll a 6 or 7 on the first roll, then roll a 6 on the second roll. The chance of a 6 or 7 on the first roll is $5/36 + 6/36 = 11/36$, so the chance of not rolling 6 or 7 is 25/36. To find the probability of winning on the second roll, we multiply this number by the probability of rolling a 6 on any individual roll, 5/36, so the probability of winning on the second roll is $(25/36)(5/36)$. To win on the third roll, the first two rolls must not be 6 or 7, then we must roll 6 on the third roll, which has probability $(25/36)(25/36)(5/36)$. The probability of winning on the fourth roll is $(25/36)^3(5/36)$, and so on. Adding all of these probabilities together, the chance of winning your bet is

$$\frac{5}{36} + \left(\frac{25}{36}\right)\left(\frac{5}{36}\right) + \left(\frac{25}{36}\right)^2\left(\frac{5}{36}\right) + \left(\frac{25}{36}\right)^3\left(\frac{5}{36}\right) + \cdots$$

$$= \frac{5}{36}\left[1 + \frac{25}{36} + \left(\frac{25}{36}\right)^2 + \left(\frac{25}{36}\right)^3 + \cdots\right]$$

$$= \frac{5}{36}\left(\frac{1}{1 - \frac{25}{36}}\right) = \frac{5}{36 - 25} = \frac{5}{11}$$

as predicted.   □

The Harmonic Series and Variations

When an infinite series adds up to a (finite) number, we say that the sum *converges* to that number. When an infinite series doesn't converge, we say that the series *diverges*. If an infinite series converges, then the individual numbers being summed need to be getting closer and closer to 0. For example, we saw that the series $1 + 1/2 + 1/4 + 1/8 + \cdots$ converged to 2, and notice that the terms $1, 1/2, 1/4, 1/8\ldots$ are getting closer and closer to 0.

But the converse statement is not true, since it is possible for a series to diverge even if the terms are heading to 0. The most important example is the *harmonic series*, so named because the ancient Greeks discovered that strings of lengths proportional to $1, 1/2, 1/3, 1/4, 1/5, \ldots$ could produce harmonious sounds.

Theorem: The harmonic series diverges. That is,

$$1 + \frac{1}{2} + \frac{1}{3} + \frac{1}{4} + \frac{1}{5} + \cdots = \infty$$

Proof: To prove that the sum is infinity, we need to show that the sum gets arbitrarily large. To do this we break up our sum into pieces based on the number of digits in the denominator. Notice that since the first 9 terms are each bigger than $1/10$, therefore

$$1 + \frac{1}{2} + \frac{1}{3} + \frac{1}{4} + \frac{1}{5} + \frac{1}{6} + \frac{1}{7} + \frac{1}{8} + \frac{1}{9} > \frac{9}{10}$$

The next 90 terms are each bigger than $1/100$, and so

$$\frac{1}{10} + \frac{1}{11} + \frac{1}{12} + \cdots + \frac{1}{99} > 90 \times \frac{1}{100} = \frac{9}{10}$$

Likewise, the next 900 terms are each bigger than $1/1000$. Thus,

$$\frac{1}{100} + \frac{1}{101} + \frac{1}{102} + \cdots + \frac{1}{999} > \frac{900}{1000} = \frac{9}{10}$$

Continuing this way, we see that

$$\frac{1}{1000} + \frac{1}{1001} + \frac{1}{1002} + \cdots + \frac{1}{9999} > \frac{9000}{10,000} = \frac{9}{10}$$

and so on. Hence the sum of all of the numbers is at least

$$\frac{9}{10} + \frac{9}{10} + \frac{9}{10} + \frac{9}{10} + \cdots$$

which grows without bound.   ☺

✕ Aside

Here's a fun fact:

$$1 + \frac{1}{2} + \frac{1}{3} + \cdots + \frac{1}{n} \approx \gamma + \ln n$$

where γ is the number $0.5772155649\ldots$ (called the Euler-Mascheroni constant) and $\ln n$ is the natural logarithm of n, described in Chapter 10. (It is not known if γ, pronounced "gamma," is rational or not.) The approximation gets better as n gets larger. Here is a table comparing the sum with the approximation.

n	$1 + \frac{1}{2} + \frac{1}{3} + \cdots + \frac{1}{n}$	$\gamma + \ln n$	Error
10	2.92897	2.87980	0.04917
100	5.18738	5.18239	0.00499
1000	7.48547	7.48497	0.00050
10,000	9.78761	9.78756	0.00005

Equally fascinating is the fact that if we only look at prime denominators, then for a large prime number p,

$$\frac{1}{2} + \frac{1}{3} + \frac{1}{5} + \frac{1}{7} + \frac{1}{11} + \frac{1}{13} + \cdots + \frac{1}{p} \approx M + \ln \ln p$$

where $M = 0.2614972\ldots$ is the *Mertens constant* and the approximation becomes more accurate as p gets larger.

One consequence of this fact is that

$$\frac{1}{2} + \frac{1}{3} + \frac{1}{5} + \frac{1}{7} + \frac{1}{11} + \frac{1}{13} + \cdots = \infty$$

but it really crawls to infinity because the log of the log of p is small, even if p is quite large. For instance, when we sum the reciprocals of all prime numbers below googol, 10^{100}, the sum is still below 6.

Let's see what happens when you modify the harmonic series. If you throw away a finite number of terms, the series still diverges. For example, if you throw away the first million terms $1 + \frac{1}{2} + \cdots + \frac{1}{10^6}$, which sums to a little more than 14, the remaining terms still sum to infinity.

If you make the terms of the harmonic series larger, then the sum still diverges. For example, since for $n > 1$, $\frac{1}{\sqrt{n}} > \frac{1}{n}$, we have

$$1 + \frac{1}{\sqrt{2}} + \frac{1}{\sqrt{3}} + \frac{1}{\sqrt{4}} + \cdots = \infty$$

But making each term *smaller* does *not* necessarily mean that the sum will converge. For example, if we divide each term in the harmonic series by 100, it still diverges, since

$$\frac{1}{100} + \frac{1}{200} + \frac{1}{300} + \cdots = \frac{1}{100}(1 + 1/2 + 1/3 + 1/4 + \cdots) = \infty$$

Yet there are changes to the series that will cause it to converge. For instance, if we square each term, the series converges. As Euler proved,

$$1 + \frac{1}{2^2} + \frac{1}{3^2} + \frac{1}{4^2} + \cdots = \frac{\pi^2}{6}$$

In fact, it can be shown (through integral calculus) that for any $p > 1$,

$$1 + \frac{1}{2^p} + \frac{1}{3^p} + \frac{1}{4^p} + \cdots$$

converges to some number below $\frac{p}{p-1}$. For example, when $p = 1.01$, even though the terms are just slightly smaller than the terms of the harmonic series, we have a convergent series

$$1 + \frac{1}{2^{1.01}} + \frac{1}{3^{1.01}} + \frac{1}{4^{1.01}} + \cdots < 101$$

Suppose we remove from the harmonic series any number with a 9 in it somewhere. In this situation, we can show that the series does not sum to infinity (and therefore must converge to something). We prove this by counting the 9-less numbers with denominators of each length. For instance, we begin with 8 fractions with one-digit denominators, namely $\frac{1}{1}$ through $\frac{1}{8}$. There are $8 \times 9 = 72$ two-digit numbers without 9, since there are 8 choices for the first digit (anything but 0 or 9) and 9 choices for the second digit. Likewise, there are $8 \times 9 \times 9$ three-digit numbers without 9s, and more generally, $8 \times 9^{n-1}$ n-digit numbers without 9s. Noting that the largest of the one-digit fractions is 1, the largest two-digit fraction is $\frac{1}{10}$, and the largest three-digit fraction is $\frac{1}{100}$, we can break our infinite series into blocks as follows,

$$1 + \frac{1}{2} + \frac{1}{3} + \frac{1}{4} + \frac{1}{5} + \frac{1}{6} + \frac{1}{7} + \frac{1}{8} < 8$$

$$\frac{1}{10} + \frac{1}{11} + \frac{1}{12} + \cdots + \frac{1}{88} < (8 \times 9) \times \frac{1}{10} = 8\left(\frac{9}{10}\right)$$

$$\frac{1}{100} + \frac{1}{101} + \frac{1}{102} + \cdots + \frac{1}{888} < (8 \times 9^2)\frac{1}{100} = 8\left(\frac{9}{10}\right)^2$$

and so on. The sum of all of the numbers is at most

$$8\left(1 + \frac{9}{10} + \left(\frac{9}{10}\right)^2 + \left(\frac{9}{10}\right)^3 + \cdots\right) = \frac{8}{1 - \frac{9}{10}} = 80$$

by the geometric series. Hence, the 9-less series converges to a number less than 80. □

One way to think of the convergence of this series is that almost all large numbers have a 9 in them somewhere. Indeed, if you generate a random number with each digit randomly chosen from 0 to 9, the chance that the number 9 was not among the first n digits would be $(9/10)^n$, which goes to 0 as n gets larger and larger.

> ✂ **Aside**
>
> If we treat the digits of π and e as random strings of digits, then it is a virtual certainty that your favorite integer appears somewhere in those numbers. For example, my favorite four-digit number, 2520, appears as digits 1845 through 1848 of π. The first 6 Fibonacci numbers 1, 1, 2, 3, 5, 8, appear beginning at digit 820,390. It's not too surprising to see this among the first million digits, since with a randomly generated number, the chance that the digits of a particular six-digit location matches your number is one in a million. So with about a million six-digit locations, your chances are pretty good. On the other hand, it is rather astonishing that the number 999999 appears so early in π, beginning at digit 763. Physicist Richard Feynman once remarked that if he memorized π to 767 decimal places, people might think that π was a rational number, since he could end his recitation with "999999 and so on."
>
> There are programs and websites that will find your favorite digit strings inside π and e. Using one of these programs, I discovered that if I memorized π to 3000 decimal places, it would end with 31961, which is amazing to me because March 19, 1961, happens to be my birthday!

Intriguing and Impossible Infinite Sums

Let's *summarize* some of the sums we have seen so far.

We began this chapter by investigating

$$1 + \frac{1}{2} + \frac{1}{4} + \frac{1}{8} + \frac{1}{16} + \cdots = 2$$

We saw that this was a special case of the geometric series, which says that for any value of x where $-1 < x < 1$,

$$1 + x + x^2 + x^3 + x^4 + \cdots = \frac{1}{1-x}$$

Notice that the geometric series also works for negative numbers between 0 and -1. For instance, when $x = -1/2$, it says

$$1 - \frac{1}{2} + \frac{1}{4} - \frac{1}{8} + \frac{1}{16} - \cdots = \frac{1}{1 - (-1/2)} = \frac{2}{3}$$

A series that alternates between positive and negative numbers that are getting closer and closer to zero is called an *alternating series*. Alternating series always converge to some number. To illustrate with the alternating series above, draw the real line and put your finger on the number 0. Then move it to the right by 1, then go to the left by $1/2$, then go to the right by $1/4$. (At this point, your finger should be on the point $3/4$.) Then go to the left by $1/8$ (so that your finger is now on the point $5/8$), and so on. Your finger will be homing in on a single number, in this case $2/3$.

Now consider the alternating series

$$1 - \frac{1}{2} + \frac{1}{3} - \frac{1}{4} + \frac{1}{5} - \frac{1}{6} + \cdots$$

After four terms, we know that the infinite sum is at least $1 - 1/2 + 1/3 - 1/4 = 7/12 = 0.583\ldots$, and after five terms, we know that it is at most $1 - 1/2 + 1/3 - 1/4 + 1/5 = 47/60 = 0.783.\ldots$ The eventual infinite sum is a little more than halfway between these two numbers, namely $0.693147.\ldots$ Using calculus, we can find the *real* value of this number.

As a warm-up exercise, take the geometric series:

$$1 + x + x^2 + x^3 + x^4 + \cdots = \frac{1}{1-x}$$

and let's see what happens when we differentiate both sides. Recall from Chapter 11 that the derivatives of 1, x, x^2, x^3, x^4, and so on are, respectively, 0, 1, $2x$, $3x^2$, $4x^3$, and so on. Thus, if we assume that the derivative of an infinite sum is the (infinite) sum of the derivatives, and use the chain rule to differentiate $(1-x)^{-1}$, then we get for $-1 < x < 1$,

$$1 + 2x + 3x^2 + 4x^3 + 5x^4 + \cdots = \frac{1}{(1-x)^2}$$

Next let's take the geometric series with x replaced by $-x$, so that for $-1 < x < 1$,

$$1 - x + x^2 - x^3 + x^4 - \cdots = \frac{1}{1 + x}$$

Now we take the *anti-derivative* of both sides, known to calculus students as *integration*. To find the anti-derivative, we go backward. For instance, the derivative of x^2 is $2x$, so going backward, we say that the anti-derivative of $2x$ is x^2. (As a technical note for calculus students, the derivative of $x^2 + 5$, or $x^2 + \pi$, or $x^2 + c$ for any number c, is also $2x$, so the anti-derivative of $2x$ is really $x^2 + c$.) The anti-derivatives of 1, x, x^2, x^3, x^4, and so on are, respectively x, $x^2/2$, $x^3/3$, $x^4/4$, $x^5/5$, and the anti-derivative of $1/(1 + x)$ is the natural logarithm of $1 + x$. That is, for $-1 < x < 1$,

$$x - \frac{x^2}{2} + \frac{x^3}{3} - \frac{x^4}{4} + \frac{x^5}{5} - \cdots = \ln(1 + x)$$

(Technical note for calculus students: the constant term on the left side is 0, since when $x = 0$, we want the left side to evaluate to $\ln 1 = 0$.) As x gets closer and closer to 1, we discover the *natural* meaning of $0.693147\ldots$, namely

$$1 - \frac{1}{2} + \frac{1}{3} - \frac{1}{4} + \frac{1}{5} - \frac{1}{6} + \cdots = \ln 2$$

>⧓ **Aside**
>
>If we write the geometric series with x replaced with $-x^2$, we get, for x between -1 and 1,
>
>$$1 - x^2 + x^4 - x^6 + x^8 - \cdots = \frac{1}{1 + x^2}$$
>
>In most calculus textbooks, it is shown that $y = \tan^{-1} x$ has derivative $y' = \frac{1}{1+x^2}$. Thus, if we take the anti-derivative of both sides (and note that $\tan^{-1} 0 = 0$), we get
>
>$$x - \frac{x^3}{3} + \frac{x^5}{5} - \frac{x^7}{7} + \frac{x^9}{9} - \cdots = \tan^{-1} x$$
>
>Letting x get closer and closer to 1, we get
>
>$$1 - \frac{1}{3} + \frac{1}{5} - \frac{1}{7} + \frac{1}{9} - \frac{1}{11} + \cdots = \tan^{-1} 1 = \frac{\pi}{4}$$

We have seen how the geometric series can be used, now let's see how it can be abused. The formula for the geometric series says that

$$1 + x + x^2 + x^3 + x^4 + \cdots = \frac{1}{1-x}$$

for values of x where $-1 < x < 1$. Let's look at what happens when $x = -1$. Then the formula would tell us that

$$1 - 1 + 1 - 1 + 1 - \cdots = \frac{1}{1-(-1)} = \frac{1}{2}$$

Of course that's impossible; since we are only adding and subtracting integers, there is no way that they should sum to a fractional value like $1/2$, even if the sum converged to something. On the other hand, the answer isn't entirely ridiculous, because when we look at the partial sums we have

$$
\begin{aligned}
1 &= 1 \\
1 - 1 &= 0 \\
1 - 1 + 1 &= 1 \\
1 - 1 + 1 - 1 &= 0
\end{aligned}
$$

and so on. Since half of the partial sums are 1 and half of the partial sums are 0, the answer $1/2$ isn't too unreasonable.

Using the illegal value $x = 2$, the geometric series says

$$1 + 2 + 4 + 8 + 16 + \cdots = \frac{1}{1-2} = -1$$

This answer looks even more ridiculous than the last sum. How can the sum of positive numbers possibly be negative? And yet maybe there is a reasonable interpretation for this sum too. For instance, in Chapter 3, we encountered ways in which a positive number could act like a negative number, with relationships like

$$10 \equiv -1 \pmod{11}$$

allowing us to make statements like $10^k \equiv (-1)^k \pmod{11}$.

Here's a way to understand $1 + 2 + 4 + 8 + 16 + \cdots$ that requires thinking out of the box a little. Recall that in Chapter 4, we observed that every positive integer can be represented as the sum of powers of 2 in a unique way. This is the basis for *binary* arithmetic, which is the way digital computers perform calculations. Every integer uses a finite number of powers of 2. For instance, $106 = 2 + 8 + 32 + 64$ uses just four powers of 2. But now suppose that we also allowed *infinite integers*, where we could use as many powers of 2 as we wanted. A typical infinite integer might look like

$$1 + 2 + 8 + 16 + 64 + 256 + 2048 + \cdots$$

with powers of 2 appearing forever. What these numbers would represent is unclear, but we could come up with consistent rules for doing arithmetic with them. For instance, we could add such numbers provided that we allowed carries to occur in the natural way. For instance, if we add 106 to the above number, we would get

$$
\begin{array}{llll}
1 + 2 & + 8 + 16 & + 64 & + 256 + \cdots \\
\underline{+ 2} & \underline{+ 8} & \underline{+ 32 + 64} & \\
1 \quad + 4 & & + 64 + 128 + 256 + \cdots
\end{array}
$$

where the $2 + 2$ combine to create the 4; next, the $8 + 8$ forms 16, but when added to the next 16 creates a 32, which when added to the next 32 creates 64, which when added to the two 64s creates 64 and 128. Everything from 256 onward is unchanged. Now imagine what happens when we take the "largest" infinite integer and add 1 to it.

$$
\begin{array}{l}
1 + 2 + 4 + 8 + 16 + 32 + 64 + 128 + 256 + \cdots \\
\underline{+1 \hspace{9cm}}
\end{array}
$$

The result would be a never-ending chain reaction of carrying with no power of 2 appearing below the line. Hence, the sum can be thought of as 0. Since $(1 + 2 + 4 + 8 + 16 + \cdots) + 1 = 0$, then subtracting 1 from both sides suggests that the infinite sum behaves like the number -1.

Here is my favorite impossible infinite sum:

$$1 + 2 + 3 + 4 + 5 + \cdots = \frac{-1}{12}$$

We "prove" this by the *shifty algebra* approach that we used in the second proof of the finite geometric series. Although the shifty approach is valid for finite sums, it can lead to nonsensical-looking results for infinite sums. For instance, let's first use shifty algebra to explain an earlier identity. We write the sum twice, but shift the terms over by one space in the second sum as follows:

$$S = 1 - 1 + 1 - 1 + 1 - 1 + \cdots$$
$$S = 1 - 1 + 1 - 1 + 1 - \cdots$$

Adding these equations together gives us

$$2S = 1$$

and therefore $S = 1/2$, as we asserted earlier, when we set $x = -1$ in the geometric series.

> ✕ **Aside**
>
> We can use the shifty algebra approach to give a quick, but not quite legal, proof of the geometric series formula.
>
> $$S = 1 + x + x^2 + x^3 + x^4 + x^5 + \cdots$$
>
> $$xS = x + x^2 + x^3 + x^4 + x^5 + \cdots$$
>
> Subtracting these two equations gives us
>
> $$S(1 - x) = 1$$
>
> and therefore
>
> $$S = \frac{1}{1 - x}$$
> ▢

Next we claim that the alternating version of our desired sum also has an interesting answer, namely

$$1 - 2 + 3 - 4 + 5 - 6 + 7 - 8 + \cdots = \frac{1}{4}$$

Here's the shifty algebra proof. Writing the sum twice, we get

$$T = 1 - 2 + 3 - 4 + 5 - 6 + 7 - 8 + \cdots$$
$$T = 1 - 2 + 3 - 4 + 5 - 6 + 7 - \cdots$$

When we add these equations we get

$$2T = 1 - 1 + 1 - 1 + 1 - 1 + 1 - 1 + \cdots$$

Therefore $2T = S = 1/2$ and so $T = 1/4$, as claimed.

Finally, let's see what happens when we write the sum of all positive integers as U and underneath that we write the previous (and unshifted) sum T.

$$U \; = \; 1+2+3+4+5+6+7+8+\cdots$$
$$T \; = \; 1-2+3-4+5-6+7-8+\cdots$$

Subtracting the second equation from the first one reveals

$$U-T = 4+8+12+16+\cdots = 4(1+2+3+4+\cdots)$$

In other words,
$$U - T = 4U$$

Solving for U, we get $3U = -T = -1/4$, and therefore

$$U = -1/12$$

as claimed.

For the record, when you add an infinite number of positive integers, the sum diverges to infinity. But before you dismiss all of these finite answers as pure magic with no redeeming qualities, it is possible that there is a context where this actually makes sense. By expanding our view of numbers, we saw a way in which the sum $1+2+4+8+16+\cdots = -1$ was not so implausible. Recall also that when we confined numbers to the real line, it was impossible to find a number with a square of -1, yet this became possible once we viewed complex numbers as occupants of the plane with their own consistent rules of arithmetic. In fact, theoretical physicists who study string theory actually use the sum $1+2+3+4+\cdots = -1/12$ result in their calculations. When you encounter paradoxical results like the sums shown here, you could just dismiss them as impossible and be done with it, but if you allow your imagination to consider the possibilities, then a consistent and beautiful system can arise.

Let's end this book with one more paradoxical result. At the beginning of this section we saw that the alternating series

$$1 - \frac{1}{2} + \frac{1}{3} - \frac{1}{4} + \frac{1}{5} - \frac{1}{6} + \cdots$$

converged to the number $\ln 2 = 0.693147\ldots$. If you add these numbers in a different order, you would naturally expect to still get the same sum, since the commutative law of addition says that

$$A + B = B + A$$

for any numbers A and B. And yet, look what happens when we rearrange the sum in the following way:

$$1 - \frac{1}{2} - \frac{1}{4} + \frac{1}{3} - \frac{1}{6} - \frac{1}{8} + \frac{1}{5} - \frac{1}{10} - \frac{1}{12} + \cdots$$

Note that these are the same numbers being summed, since every fraction with an odd denominator is being added and every fraction with an even denominator is being subtracted. Even though the even numbers are being used up at twice as fast a rate as the odd numbers, they both have an inexhaustible supply, and every fraction from the original sum appears exactly once in the new sum. Agreed? But notice that this equals

$$= \left(1 - \frac{1}{2}\right) - \frac{1}{4} + \left(\frac{1}{3} - \frac{1}{6}\right) - \frac{1}{8} + \left(\frac{1}{5} - \frac{1}{10}\right) - \frac{1}{12} + \cdots$$

$$= \frac{1}{2} \qquad - \frac{1}{4} \qquad + \frac{1}{6} \qquad - \frac{1}{8} \qquad + \frac{1}{10} \qquad - \frac{1}{12} + \cdots$$

$$= \frac{1}{2}\left(1 \qquad - \frac{1}{2} \qquad + \frac{1}{3} \qquad - \frac{1}{4} \qquad + \frac{1}{5} \qquad - \frac{1}{6} + \cdots\right)$$

which is one-half of the original sum! How can this be? How is it possible that when we rearrange a collection of numbers, we can get a completely different number? The surprising answer is that the commutative law of addition can actually *fail* when you are adding an infinite number of numbers.

This problem arises in a convergent series whenever both the positive terms and the negative terms form divergent series. In other words, the positive terms add to ∞ and the negative terms add to $-\infty$. Such was the case with our last example. These sequences are called *conditionally convergent* series and, amazingly, they can be rearranged to obtain any total you desire. How would we rearrange the last sum to get 42? You would add enough positive terms until the sum just exceeds 42, then subtract your first negative term. Then add more positive terms until it exceeds 42 again. Then subtract your second negative term. Repeating this process, your sum will eventually get closer and closer to 42. (For instance, after subtracting your fifth negative term, $-1/10$, you will always be within 0.1 of 42. After subtracting the fiftieth negative term, $-1/100$, you will always be within 0.01 of 42, and so on.)

Most of the infinite series that we encounter in practice do not exhibit this strange sort of behavior. If we replace each term with its *absolute value* (so that each negative term is turned positive), then if that new sum converges, then the original series is called *absolutely convergent*. For example, the alternating series we encountered earlier,

$$1 - \frac{1}{2} + \frac{1}{4} - \frac{1}{8} + \frac{1}{16} - \cdots = \frac{2}{3}$$

is absolutely convergent, since when we sum the absolute values we get the familiar convergent series

$$1 + \frac{1}{2} + \frac{1}{4} + \frac{1}{8} + \frac{1}{16} + \cdots = 2$$

With absolutely convergent series, the commutative law of addition will always work, even with infinitely many terms. Thus in the alternating series above, no matter how thoroughly you rearrange the numbers 1, $-1/2, 1/4, -1/8\ldots$, the rearranged sum will always converge to 2/3.

Unlike an infinite series, a book has to end sometime. We don't dare to try to go beyond infinity, so this seems like a good place to stop, but I can't resist one last mathemagical excursion.

Encore! Magic Squares!

As a reward for making it all the way to the end of the book, here is one more magical mathematical topic for your enjoyment. It has nothing to do with infinity, but it does have the word "magic" squarely in its title: magic squares. A *magic square* is a square grid of numbers where every row, column, and diagonal add to the same number. The most famous 3-by-3 magic square is shown below, where all three rows, all three columns, and both diagonals add to 15.

4	9	2
3	5	7
8	1	6

A 3-by-3 magic square with magic total 15

Here's a little-known fact about this magic square that I call the square-palindromic property. If you treat each row and column as a

3-digit number and take the sum of their squares, you will find that

$$492^2 + 357^2 + 816^2 = 294^2 + 753^2 + 618^2$$

$$438^2 + 951^2 + 276^2 = 834^2 + 159^2 + 672^2$$

A similar phenomenon occurs with some of the "wrapped" diagonals, too. For instance,

$$456^2 + 312^2 + 897^2 = 654^2 + 213^2 + 798^2$$

Magic "squares" indeed!

The simplest 4-by-4 magic square uses the numbers 1 through 16 where all rows, columns, and diagonals sum to the magic total of 34, like the one below. Mathematicians and magicians like 4-by-4 magic squares because they usually contain dozens of different ways to achieve the magic total. For instance, in the magic square below, every row, column, and diagonal adds to 34, as does every 2-by-2 square inside it, including the upper left quadrant $(8, 11, 13, 2)$, the four numbers in the middle, and the four corners of the magic square. Even the wrapped diagonals sum to 34, as do the corners of any 3-by-3 square inside.

8	11	14	1
13	2	7	12
3	16	9	6
10	5	4	15

A magic square with total 34. Every row, column, and diagonal sums to 34, as do nearly every other symmetrically placed four squares.

Do you have a favorite two-digit number bigger than 20? You can instantly create a magic square with total T just by using the numbers 1 though 12, along with the four numbers $T - 18$, $T - 19$, $T - 20$ and $T - 21$ as shown on the next page.

For example, see the magic square on the next page with a magic total of $T = 55$. Every group of four that used to add up to 34 will now add up to 55, as long as the group of 4 includes exactly one (not two, not zero) of the squares that use the variable T. So the upper right squares will have the correct total ($35 + 1 + 7 + 12 = 55$) but the middle left squares will not ($34 + 2 + 3 + 37 \neq 55$).

8	11	$T - 20$	1
$T - 21$	2	7	12
3	$T - 18$	9	6
10	5	4	$T - 19$

A quick magic square with magic total T

8	11	35	1
34	2	7	12
3	37	9	6
10	5	4	36

A magic square with total 55

Although not everyone has a favorite two-digit number, everyone does have a birthday, and I find that people appreciate personalized magic squares that use their birthdays. Here is a method that I use for creating a "double birthday" magic square, where the birthday actually appears twice: in the top row and in the four corners. If the birthday uses the numbers A, B, C, and D, then you can create the following magic square. Notice that every row, column, and diagonal, and most symmetrically placed groups of four squares, will add to the magic total $A + B + C + D$.

A	B	C	D
$C - 1$	$D + 1$	$A - 1$	$B + 1$
$D + 1$	$C + 1$	$B - 1$	$A - 1$
B	$A - 2$	$D + 2$	C

A double birthday magic square. The date A/B/C/D appears in the top row and four corners.

For my mother's birthday, November 18, 1936, the magic square looks like this:

11	18	3	6
2	7	10	19
7	4	17	10
18	9	8	3

A birthday magic square for my mother: 11/18/36, with magic total 38

Now create a magic square based on your own birthday. If you follow the pattern given above, your birthday total will appear more than three dozen times. See how many you can find.

Although 4-by-4 magic squares have the most combinations, there are techniques for creating magic squares of higher order. For example, here is a 10-by-10 magic square using all the numbers from 1 to 100.

92	99	1	8	15	67	74	51	58	40
98	80	7	14	16	73	55	57	64	41
79	6	88	20	22	54	56	63	70	47
85	87	19	21	3	60	62	69	71	28
86	93	25	2	9	61	68	75	52	34
17	24	76	83	90	42	49	26	33	65
23	5	82	89	91	48	30	32	39	66
4	81	13	95	97	29	31	38	45	72
10	12	94	96	78	35	37	44	46	53
11	18	100	77	84	36	43	50	27	59

A 10-by-10 magic square using numbers 1 through 100

Can you figure out the magic total of each row, column, and diagonal without adding up any of the rows? Sure! Since we showed a long time ago that numbers 1 through 100 add to 5050, each row must add to one-tenth of that. Hence the magic total must be $5050/10 = 505$. This book began with the problem of adding the numbers from 1 to 100, and so it seems appropriate that we end here as well. Congratulations (and thank you) for reading to the end of the book. We covered a great many mathematical topics, ideas, and problem-solving strategies. As you go back through this book and read other books that rely on mathematical thinking, I hope you find the ideas presented in this book to be useful, interesting, and magical.

Aftermath

I hope that this is not the last math book you ever read, since there is so much good material out there. Indeed, most of the really interesting math that I have learned came from outside of the classroom, and included many of the items listed here.

This book is an outgrowth of my video course *The Joy of Mathematics*, produced by The Great Courses. That course contains twenty-four 30-minute lectures by me on all of the topics covered in this book, plus a few more topics, like The Joy of Probability, Mathematical Games, and Magic. (I am grateful for their willingness to let me borrow many ideas from that course for this book.) The Great Courses has more than three dozen mathematics courses (available on audio, video, and downloadable formats) on many mathematical topics, including entire courses devoted to topics like algebra, geometry, calculus, and history of mathematics. They have done an excellent job at finding some of the best professors in the country to teach these courses, and it has been an honor and a privilege to create four courses for them. My other three courses are *Discrete Mathematics*, *The Secrets of Mental Math*, and *The Mathematics of Games and Puzzles*.

For printed information on doing math in your head, see my book *Secrets of Mental Math*, written with Michael Shermer, published by Random House. It goes into great detail on how to quickly and accurately answer all kinds of problems, big and small. If you know your multiplication tables through 10, then you should be able to understand all of the techniques in the book. For a more elementary approach, I have created a workbook on doing mental addition and subtraction that is aimed at elementary-school-age children, called *The Art of Mental Calculation* (co-authored and beautifully illustrated by Natalya St. Clair). You can find it on Amazon.com or createspace.com.

I have written three other mathematics books for advanced readers. The Mathematical Association of America (MAA) has published *Proofs*

That Really Count: The Art of Combinatorial Proof, with Jennifer J. Quinn, and *Biscuits of Number Theory*, co-edited with Ezra Brown. My most recent book is *The Fascinating World of Graph Theory* with Gary Chartrand and Ping Zhang, published by Princeton University Press.

I owe a literary debt of gratitude to Martin Gardner, the greatest mathemagician of all time, who wrote more than two hundred books, many on the subject of recreational mathematics. His books (and "Mathematical Games" columns in *Scientific American*) inspired many generations of mathematicians and math enthusiasts. Following in Gardner's footsteps, I also recommend all books by Alex Bellos, Ivars Peterson, and Ian Stewart. One of the best new books in this genre is *The Joy of X: A Guided Tour of Math, from One to Infinity* by Steven Strogatz.

For mathematics textbooks aimed at the high end of the math ability spectrum, I am a huge fan of *The Art of Problem Solving* book series, by Richard Rusczyk. These include challenging but clearly written books on algebra, geometry, calculus, problem solving, and more. Their website (ArtOfProblemSolving.com) also offers online courses for students who enjoy mathematics and participate in math contests.

There are other enjoyable online resources too. My colleague Francis Su has hundreds of examples of amazing mathematics on his Math Fun Facts page (www.math.hmc.edu/funfacts). They were originally designed for teachers who wanted to do something quick and interesting in the first five minutes of class. Alex Bogomolny has created a website, Cut the Knot (Cut-The-Knot.org) with dozens of "Interactive Mathematics Miscellany and Puzzles" that will keep you entertained for a long time. One of his columns provides over one hundred proofs of the Pythagorean theorem. For a fun, free video resource, check out the videos created by Numberphile (Numberphile.com), which present mathematics in a most entertaining way.

I have nothing more to add (or multiply), so happy reading!

Acknowledgments

This book would not have existed without the persistent encouragement of my literary agent, Karen Gantz Zahler, and the enthusiastic support from my extraordinary editor, TJ Kelleher, from Basic Books.

I cannot imagine how I could have ever completed this project without the invaluable assistance of Natalya St. Clair, who created the many figures, illustrations, and mathematical diagrams that adorn this book. Natalya has a gift for making mathematics look beautiful and is a joy to work with.

I received tons of valuable feedback from my former student Sam Gutekunst, who carefully read every chapter of this book. Sam improved the book in so many ways that he made TJ's job considerably easier. I was also fortunate to have the keen eyes of mathematicians Amy Shell-Gellasch and Vincent Matsko, who read every chapter and offered many suggestions that significantly impacted the final product.

I am fortunate to have so many wonderful colleagues and students at Harvey Mudd College. Special thanks to Professor Francis Su for many conversations and his Math Fun Facts website, and Scott and Carol Ann Smallwood for endowing the Smallwood Family Chair in Mathematics. I am also grateful to Christopher Brown, Gary Chartrand, Jay Cordes, John Fort, Ron Graham, Mohamed Omar, Jason Rosenhouse, and Natalya St. Clair for valuable discussions and ideas.

I am grateful to Ethan Brown for sharing his mnemonics for memorizing τ, Doug Dunham for permission to use his butterfly image, Dale Gerdemann for creating the Sierpinski diagram, Mike Keith for permission to use his great π tribute, "Near a Raven," mathemusicians Larry Lesser and Dane Camp for permission to use their lyrics to "Mathematical Pi" and "Knowin' Induction," and Natalya St. Clair for the "Golden Rose" photo.

Thanks to the very professional staff at Perseus Books. It was a pleasure to work with Quynh Do, TJ Kelleher, Cassie Nelson, Melissa

Veronesi, Sue Warga, and Jeff Williams (along with countless others who worked behind the scenes).

I owe an enormous thank you to The Great Courses for producing such wonderful DVD courses that have allowed me to bring mathematics to the masses in a way that I never thought possible, and for permitting me to extensively borrow material from my *Joy of Mathematics* course when preparing this book. Throughout all of these courses, Jay Tate has been an indispensable resource.

I thank my wonderful parents, Larry and Lenore Benjamin, and the teachers who shaped who I am today. I shall be forever grateful to elementary school teachers Betty Gold, Mary Ann Sparks, Jean Fisler, and the students and faculty of the math and applied math departments at Mayfield High School, Carnegie Mellon University, Johns Hopkins University, and Harvey Mudd College.

And most of all, I thank my wife, Deena, and daughters, Laurel and Ariel, for their love and patience as this book was being written. Deena proofread everything I wrote and has my eternal love and gratitude. Thank you Deena, Laurel, and Ariel for adding so much magic to my life.

Arthur Benjamin
Claremont, California
2015

Index

ARTHUR BENJAMIN holds a PhD from Johns Hopkins University and is the Smallwood Family Professor of Mathematics at Harvey Mudd College. Dr. Benjamin has received numerous awards for his writing and teaching, and served as editor of *Math Horizons* magazine for the Mathematical Association of America. He has given three TED talks, which have been viewed over 10 million times. *Reader's Digest* calls him "America's best math whiz." The author of *Secrets of Mental Math*, he lives in Claremont, California, with his wife and two daughters.

6